The Tree Identification Book

THE
Tree Identification
BOOK

A New Method for the Practical Identification and Recognition of Trees

By

GEORGE W. D. SYMONDS

Photographs by Stephen V. Chelminski

WILLIAM MORROW & COMPANY, INC. · New York

Also by George W. D. Symonds

with photographs by A. W. Merwin

THE SHRUB IDENTIFICATION BOOK

Published simultaneously in the Dominion of Canada by George J. McLeod Limited, Toronto.
Printed in the United States of America.
Library of Congress Catalog Card No. 58-5359

. . . . .

To:

M. P. S.

ACKNOWLEDGMENTS

Stephen V. Chelminski, who did all the photography for this book, was more than a fine photographer. Every item photographed was carefully discussed between us with its ultimate use in mind. He was always interested in the book and constantly on the lookout for trees, locating a number of hard-to-find specimens.

I particularly want to thank Mr. E. J. Alexander of the New York Botanical Garden for his generous assistance and encouragement. I also want to express my thanks to Mr. L. P. Politi, head gardener of the New York Botanical Garden, whose extensive knowledge of the plants and their location in the garden was invaluable in collecting specimens used in the book. His assistant, Miss Bride McSweeney, was extremely helpful in looking up locations of specific trees in their files.

I also would like to acknowledge the kind assistance of the following:

Mr. George Kalmbacher, plant identification specialist, of the Brooklyn Botanic Garden

Dr. Jerry S. Olson, forest ecologist, at the Connecticut Agricultural Experiment Station, New Haven, Connecticut

Dr. Donald Wyman, Horticulturist, Arnold Arboretum, Jamaica Plain, Boston, Massachusetts

Dr. Francis de Vos, United States National Arboretum, Washington, D. C.

Dr. H. R. Totten, University of North Carolina, Chapel Hill, North Carolina

Dr. Hugo L. Blomquist, Professor Emeritus of Botany, Durham, North Carolina

The use of material from the following is gratefully acknowledged:

The New York Botanical Garden, Bronx, New York
The Brooklyn Botanic Garden, Brooklyn, New York
The Arnold Arboretum, Jamaica Plain, Boston, Massachusetts
The United States National Arboretum, Washington, D. C.

CONTENTS

OPPOSITE—OPP
THORNS—TH
LEAVES—LF
FLOWERS—FL
FRUIT—FR
TWIGS & BUDS—TW
BARK—BK
NEEDLE—NE
MASTER PAGES—MP

INTRODUCTION

The aim of this book is to present, visually, details of trees essential for practical identification, which in turn leads to tree recognition. The distinction between tree identification and tree recognition should be clearly understood at the outset. Identification is based on observation of details. Recognition means knowing trees at a glance, just as one recognizes one's friends.

Birds conform to a pattern—two legs, two wings, definite markings—but not so trees. There are certain advantages in tree identification, however, which the bird watcher doesn't have. A tree does not become alarmed and fly away. A tree has parts which can be touched at will, and all the details pictured in the Keys of this book, with the exception of the bark, can easily be collected and carried away. It is not necessary to take the book into the field to identify most trees, and if a second look is necessary, the tree can usually be revisited, with or without the book.

The botanist is required to know methods of precise identification based on minute observation, often requiring the use of a magnifying glass or microscope. But practical identification is based on details clearly seen by the eye.

An important fact to appreciate is that precise identification is not always possible and sometimes is a matter of opinion based on weighing the relative importance of conflicting facts. For example, a Black Oak can be described as having ten or more specific features which, together, identify a tree as a Black Oak. However, in practice, one seldom finds an Oak that has all these features. It is more likely that a particular tree will have only some of them, while other details will be those characteristic of another Oak. It is thus apparent that only by deciding which features seem most important can the tree be placed in a somewhat dubious category. This responsibility rests with the botanist, and there are often honest differences of opinion on these matters.

This curious situation should not deter the amateur—on the contrary, it simplifies his problem. Precise identification is not only debatable but totally unnecessary for a full enjoyment of trees. Identification sufficient for practical purposes is possible for almost all kinds of trees by following the methods set forth in this book. Further, it will be found that observing trees, identified in this way, through the seasons and under varying conditions leads to immediate recognition, which is all that many lumbermen, foresters or nurserymen find necessary. For those who are stimulated to pursue the subject further, an appendix is included which gives an introduction to the more detailed study of botanical methods and the use of botanical manuals.

Identification actually means dividing the plant kingdom into botanical groups, starting with large, generally related groups, and subdividing into smaller, more closely related ones. Genus (plural genera) is the division represented by the main name of a tree, such as the genus Oak, genus Maple, etc. Species qualifies the genus, as in the various species of the genus Maple: Sugar Maple, Silver Maple or Red

Maple. The botanist places the genus name first with a capital letter—Acer (Maple), followed by a specific epithet in small letters—rubrum (red), the two together forming the species name—Acer rubrum (Red Maple). Varieties subdivide the species still further, the variety name consisting of the species name followed by a qualifying word, as in Acer rubrum columnare (Columnar Red Maple). The scientific names are useful because they are understood in all countries, whereas several popular names may be applied to the same tree. For example, the scientific name for a certain tree is *Nyssa sylvatica,* but it is called locally a Black Gum, Sour Gum, Tupelo, Black Tupelo, Pepperidge or even Beetle Bung Tree.

The various genera are fairly distinct and less subject to differences of opinion. The species, also, are usually clear-cut, but it is at this point that the above-mentioned difficulties begin to appear. When one gets into varieties, real problems arise, and therefore it is suggested that until further acquaintance tempts one into these difficult waters, the beginner would do well to stop at identifying genera and many of the species.

Thus the main principle of identification is elimination: vast numbers reduced to workable groups, and selection, resulting in identification, coming within these small groups. The small group is the genus; the final selection, in this case, the species. Once the genus is known, trees not included here—such as local types, trees outside the region covered, and even foreign trees—can be identified by consulting a good book or manual for any particular area. The primary difficulty ordinarily encountered will no longer exist, and instead of an aimless search among endless possibilities, an intelligent, directed approach is substituted.

The trees described in this book grow in a region roughly bounded by Maine (extending into Canada), west to North Dakota and south into Texas and along the Gulf of Mexico to northern Florida. Of course, not every tree is found in all parts of this area, and, as this book is not a complete tree manual, certain trees were of necessity omitted. Such omissions include:

(1) Strictly southern or subtropical trees, or those of another region which overlaps small sections of the area covered.

(2) Some trees of genera whose species are very numerous. It has been felt in these cases that a number of the most important and representative species would be sufficient and would enable anyone interested in further species identification to do so in a botanical manual. The Oaks are an example of this, for with over fifty species in this region alone, it is obviously impractical to include them all.

(3) Hybrids, varieties and foreign trees. However, a number of foreign trees which have become naturalized in this country are included.

The trees in this country have been restricted, from time to time, to certain areas by a combination of temperature, moisture, soil conditions and other factors, some of which are unknown or only partially understood. Some of these restrictions were imposed in the distant past, and trees that thousands of years ago grew over a wide area are now found only in a few places. The Sequoias and Redwoods are examples of this. The expressions "natural range" and "native trees," therefore, are meaningful concepts only if applied to areas occupied by certain trees at a specific geologic date.

With the white man came not only foreign trees but the desire to transplant, thereby introducing new factors into the natural-selection process of trees in this country. A number of foreign trees planted here have "escaped" and have become naturalized over wide areas. Conversely, American trees have moved outside the areas where they were found by the first settlers. It is true that at present the trees in the wilderness areas have not felt the full impact of the above forces. However, it is apparent that the "original ranges of native trees" can be used only with decreasing accuracy as an aid to identification. Thus, although range maps are interesting, per se, and useful for other reasons, they have been omitted from this book. In cases where it is helpful to know the northern or southern limits tolerated by certain trees, this is indicated and also some indication is given of the areas where the trees are most plentiful. However, this is done with the warning that trees may be found outside their theoretical limits, particularly near cities, in parks and botanical gardens.

The botanical names in this book follow Alfred Rehder's, *Manual of Cultivated Trees and Shrubs,* Macmillan, second edition. At least one, and in some cases several, common names are given for each tree, with an apology for the omission of someone's favorite.

The arrangement of the genera in the Master Pages is based on botanical divisions beginning with phyla, the largest one, and then by classes, orders and families down to genera. These larger divisions are not indicated by name, as they are not necessary for our purpose, but it is well to know that they exist and that closely related genera will be found near each other in the Master Pages.

HOW TO USE THIS BOOK

Tree identification is based on definite characteristics. No two trees have all the same characteristics, thus by finding a combination of specific details which apply to only one tree, all other trees are eliminated. This is done quickly and simply in this book by using photographs which show these characteristics.

It is important to realize and to keep firmly in mind that, for practical purposes, tree names are based on two botanical divisions: (1) the genus (Oak, Maple, etc.) and (2) the species (the kind of Oak or Maple, such as Red Oak, Black Oak, Sugar Maple or Red Maple.) This book, therefore, is divided into two main parts: the first, which is called the Pictorial Keys, is designed for genus identification, and the second, called Master Pages, is for species identification.

The Pictorial Keys, for genus identification, group such things as leaves, flowers, fruit, twigs and bark in separate sections, each forming a Key. Within each Key, things that look alike are placed together. This accomplishes two things: (1) any given detail can be found quickly and (2) similar details seen next to each other can be compared and their differences easily noted. The common name given under the Key pictures often includes the species as well as the genus name, as in some instances the species is representative of the genus. Do not, therefore, use the names in the Keys for final species identification.

The Pictorial Keys are arranged in the order of their most efficient use, and until familiarity with the method makes the order optional, it is best to begin with Key #1 and go through each Key so that no detail will be missed. Do not, for example, start with Key #7, Bark, as too many possibilities may be found, whereas the Bark Key will prove invaluable in choosing between a few possibilities selected in previous Keys. Also, do not assume that the Leaf or Fruit Keys will not be useful in winter. Very often fruit still on the tree or on the ground will prove to be the deciding clue. Material under the tree should be used with caution, of course, as wind or animals may collect false evidence. However, a dead Poplar leaf under a tree known to be a Poplar may tell which kind of Poplar it is.

Under each picture of a specific detail in the Keys is a number. This number refers to a Master Page which brings together all the details of any given tree, including a picture of the whole tree. Whenever a particular genus, such as Oak, includes two or more species, these are shown together, and final identification is made by contrasting their different details. The same details are not always used to separate one species from another. For example, the Pignut and Bitternut Hickories can be told apart by their buds, although their barks may appear to be similar, whereas the Pignut and Shagbark Hickories can be distinguished from each other by their barks, but not always by their buds.

Individual trees of the same species often vary more among themselves in *some* details than they do from other species. However, the distinguishing details for a given species remain relatively constant and can be depended on for purposes of identification. In a case where this is not found to be so, the tree is probably a cross or hybrid, something that frequently occurs among Oaks, Willows, Poplars, Hickories and Ashes. Do not let this be a source of discouragement, but rather realize that this is a fact that is easy to see and understand. If one knows that a certain tree tends to look like a White Willow but has certain features of a Crack Willow, it is more interesting, and one knows more about the tree than a person who believes that a tree must be *all* White Willow or *all* Crack Willow.

The pictures in this book are, where possible, actual size. When not, the size is indicated, and similar details are always compared in the same scale.

Each Key has a short introduction explaining how to use it and the things to look for. It is a good idea to read these through before using the Key for the first time. Thereafter, they can be referred to when a tree is difficult to identify, in case some detail has been overlooked.

Each page is indexed on the edge, but when the book is used extensively, it is suggested that index tabs be attached at the beginning of each Key.

A pocket magnifying glass will prove useful and entertaining but is not essential.

This book is designed for tree identification based on details, but do not fail, once a tree is so identified, to observe it under various conditions and at different seasons. Notice the look of the tree as a whole, how it branches, the size of the twigs silhouetted against a winter sky, the way the leaves are placed and how they hang in summer, the coloring in the fall. Realize that large twigs usually support heavy compound or large simple leaves. The reverse is also true. Try to see how the general outline of the tree differs from all other trees. The Master Pages will be helpful in pointing out what to look for. In this way a "feel" for the tree is built up, and it is only then that one can recognize a tree at a distance, such as a White Oak, whether one hundred feet tall or only thirty, growing massive and spreading in the open, or straight and tall in the forest.

PART I

PICTORIAL KEYS

BROAD-LEAVED TREES

ALL ARE DECIDUOUS (losing leaves in fall) except Holly which is evergreen.

CONTENTS:

Key #1 **OPPOSITE**—Leaves, Buds and Branching

Maple...*Acer*
Ash...*Fraxinus*
Paulownia.......................................*Paulownia*
Catalpa..*Catalpa*
Buckeye...*Aesculus*
  Horsechestnut...........................*Aesculus hippocastanum*
Nannyberry.....................................*Viburnum lentago* (also all other *Viburnums*)
Flowering Dogwood..............*Cornus florida* (Also all other Dogwoods except Alternate-leaved Dogwood—*Cornus alternifolia*)

The leaves of these seven trees (genera) grow opposite *along* the twig, those of all other deciduous trees included in this book are alternate. It is, therefore, an important characteristic to look for. The Maples and Ashes, especially, are very common, and this feature alone often leads to their identification. Note that the Paulownia and Catalpa have very similar leaves, but the Paulownia is typically opposite, whereas the Catalpa, instead of growing leaves in pairs, sends out three leaves at a point (node) around the twig in what is called a whorl, but the leaves are not alternate. Even in winter, the opposite-leaf characteristic is still plainly evident, as the leaf scars (where the leaves were attached to the twig before falling) are clearly opposite. The Catalpa, of course, will show a whorl of three leaf scars around the twig.

The Maples, Ashes and Viburnums also have buds, twigs and small branches growing opposite. Notice that the Maple twigs are more profuse and smaller than those of the Ash, which has relatively few and rather large twigs. Look up from under the trees to see this. In winter this feature will prove very helpful in the woods where the trees are high-branched and often out of reach.

Dogwoods usually have pronounced opposite branching. The Horsechestnut and other Buckeyes sometimes do, but are not consistently so. The Paulownia and Catalpa usually do *not* branch opposite. This results from the lack of development of twigs on one side of the branch, which actually should produce opposite growth.

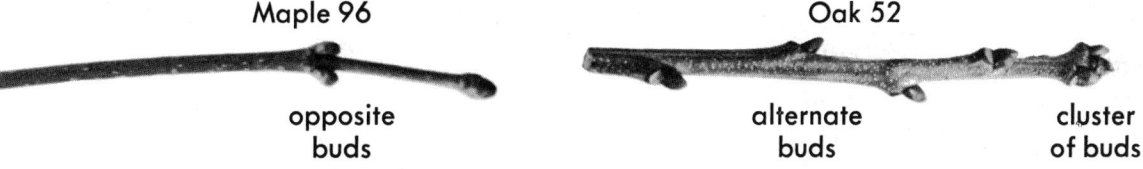

Other trees sometimes appear to branch opposite but actually do not. The Oaks, particularly, give this impression. This is caused by a cluster of growth buds at the tip of a twig, and not by growth coming from opposite buds along a twig. A glance at an Oak twig and a Maple twig will show this at once.

Maple 96

Oak 52

SCALE
½ ACTUAL SIZE

Maple 96

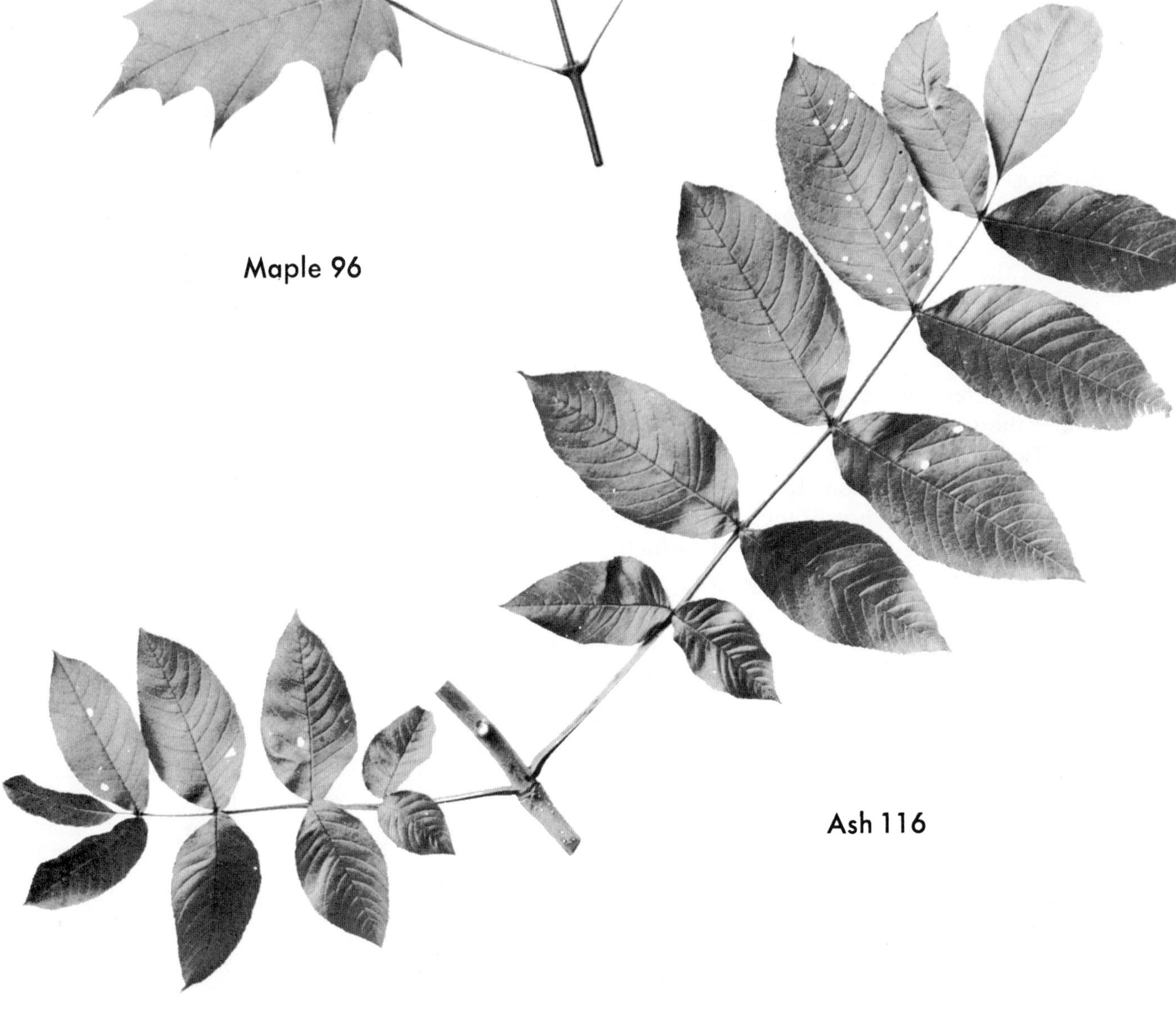

Ash 116

OPPOSITE—Leaves

Note: Catalpa usually produces leaves in whorls of three around the stem, but definitely is *not* alternate.

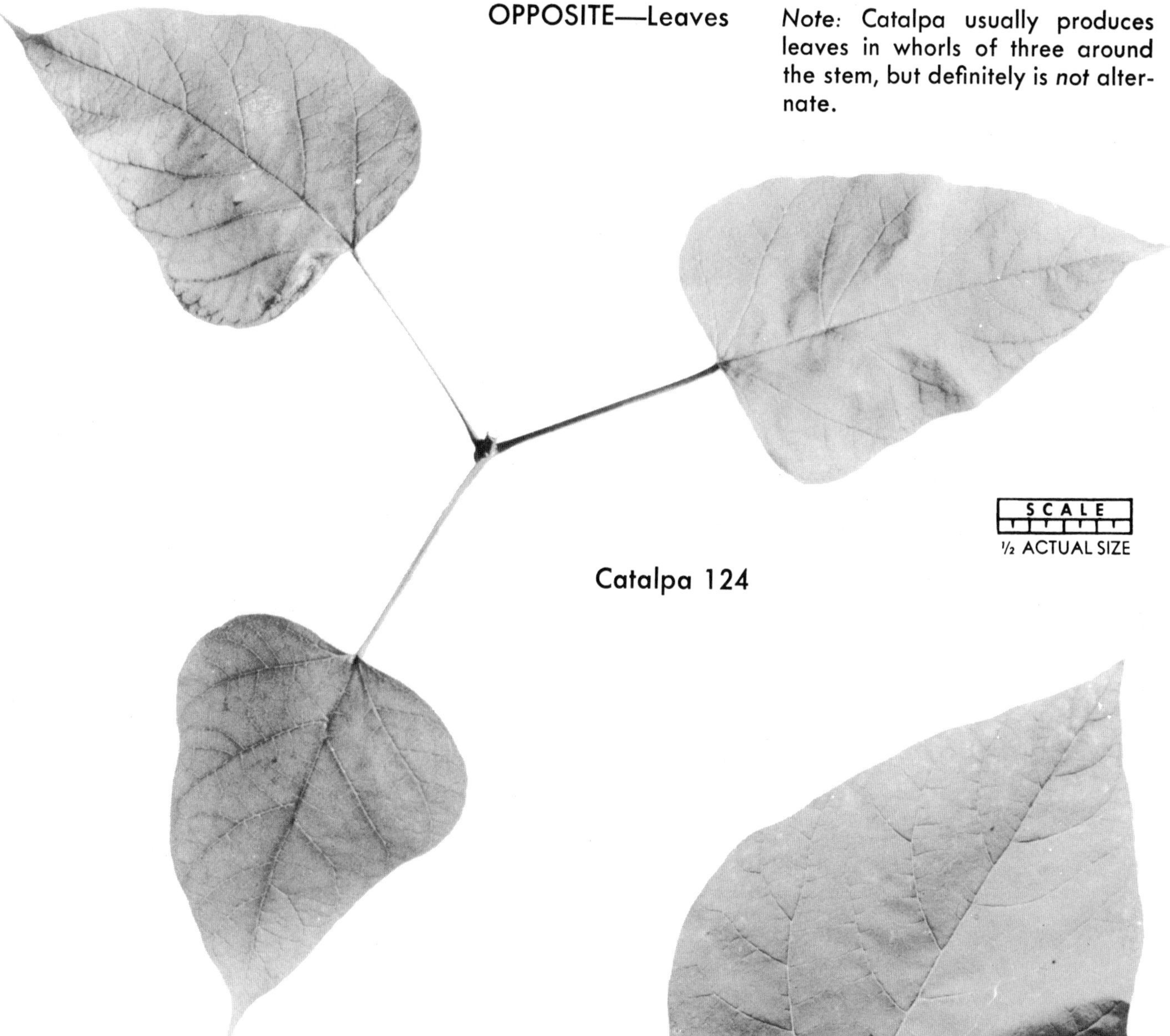

SCALE
½ ACTUAL SIZE

Catalpa 124

Paulownia 122

SCALE
½ ACTUAL SIZE

Buckeye 106

Flowering Dogwood 112

SCALE
½ ACTUAL SIZE

Horsechestnut 106

Nannyberry 127

OPPOSITE—Buds, Branching as seen in winter

SCALE
ACTUAL SIZE

Ash 116

Maple 96

Horsechestnut
sticky buds

Horsechestnut 106

Buckeye
buds not sticky

Buckeye 106

OPPOSITE—Buds, Branching as seen in winter

SCALE
ACTUAL SIZE

Note: Opposite buds are apparent on all but Dogwood, Catalpa and Paulownia. Opposite branching is seen on all but Catalpa and Paulownia.

Nannyberry 127

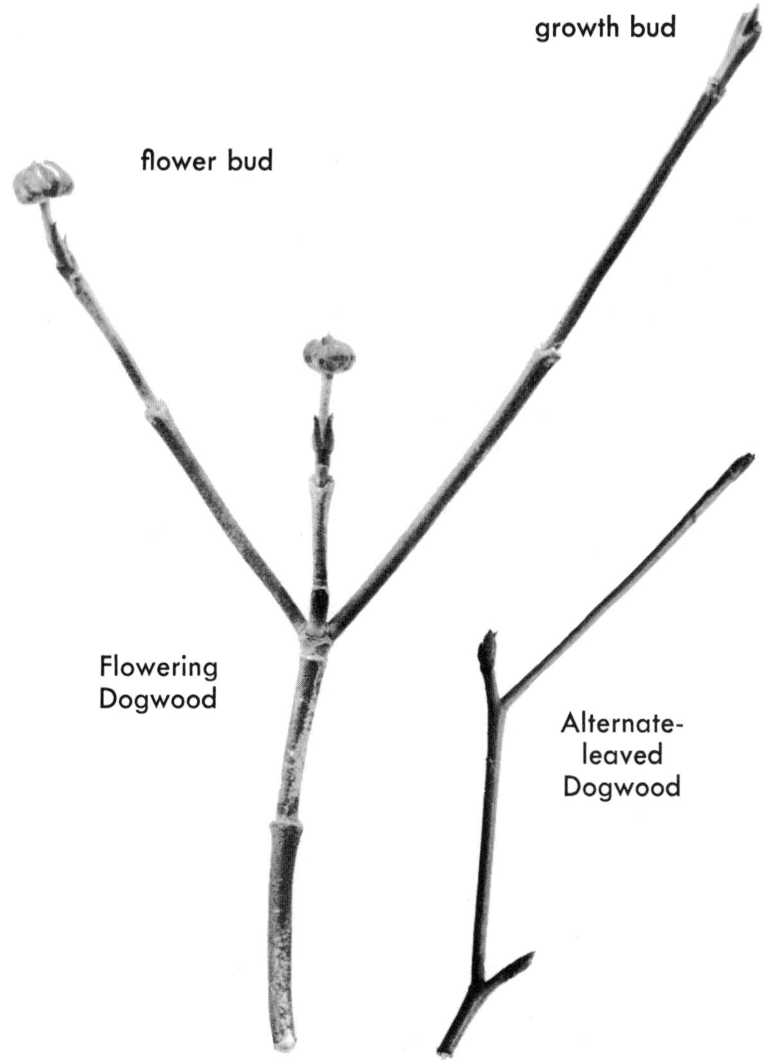

Dogwood 112

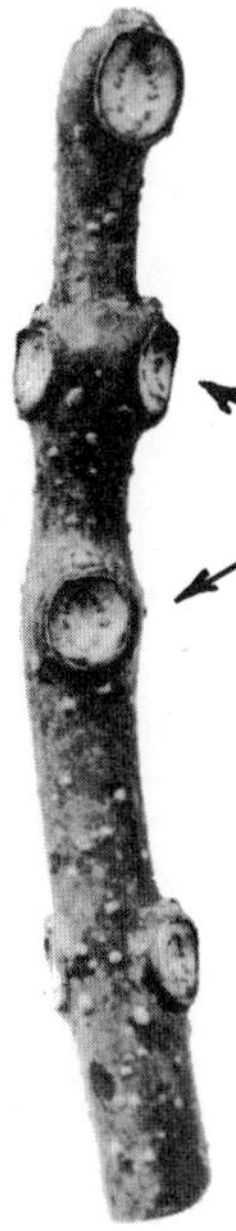

Catalpa 124

Paulownia 122

Note: All Dogwoods are opposite-leaved and -branching except Alternate-leaved Dogwood shown above, next to the Flowering Dogwood twig.

Key #2 **THORNS**—All actual size

Hawthorn
Osage Orange
Black Locust
Honey Locust

Note: Wild Plum and Buckthorn, small trees or shrubs have thorny growths, but are not included in this book.

SCALE
ACTUAL SIZE

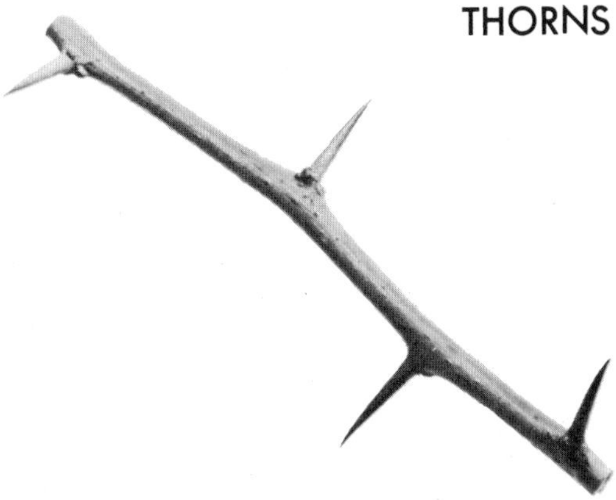

Osage Orange 62

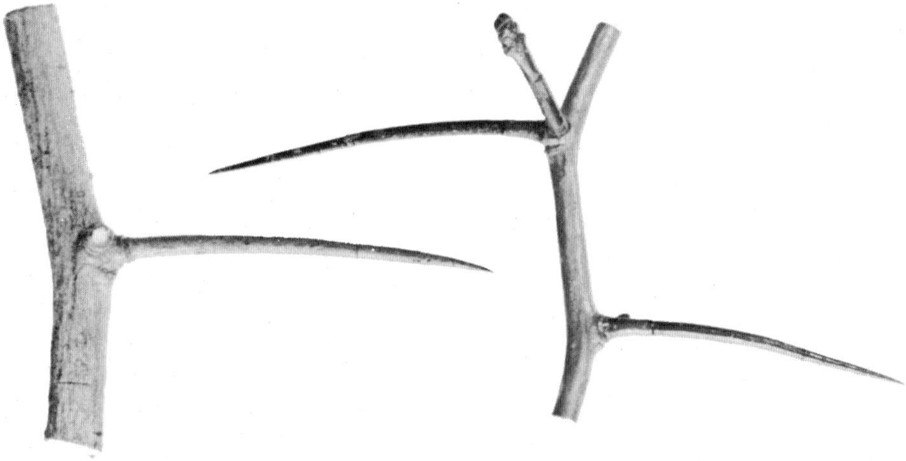

Hawthorn 73

Common or
Black Locust 81

Honey Locust 82

Key #3 **LEAVES**

This Key is divided into two major parts:

Simple Leaves—one main leaf blade on a stem.

Compound leaves—more than one leaflet on a stem, the whole making up the leaf. This sometimes drops as a unit in the fall; in any event, all the parts, including the stem, eventually drop off.

Simple Leaves

- Oak leaves
- Maple and similar leaves, including Sycamore, Sweet Gum and Tulip Tree.
- Miscellaneous. This is a large group, and the leaves are arranged for ease of selection:
  - They range from wide to narrow.
  - Where possible, leaves with smooth margins are placed at the top of the page, those with large teeth along the margins are at the bottom of the page.
  - Note: Odd-shaped and very large simple leaves, which do not seem to fit in with the above, are grouped together at the end of this subsection.

Compound Leaves

- Palmately compound leaves: leaflets coming from a single point on the stem. Example: Horsechestnut.
- Pinnately compound leaves: leaflets ranging along both sides of the stem. Example: Pecan.

Leaves vary greatly in size, compound leaves of Ailanthus, for example, ranging from less than a foot to three feet or more. Leaves vary greatly in shape also, even on the same tree. Those of Sassafras, Mulberry and Oak are particularly variable. Therefore, always look for a typical leaf or a variety of types if they appear on the same tree. Never be satisfied with the first leaf within reach.

Different species of Oaks often have similar leaves, so don't depend on the leaf to separate one Oak from another. Oak leaves are distinctive, however, and are very useful in telling an Oak from other trees. There is one exception to this—the Willow Oak group. The Willow Oak leaf will not be found in the Oak Section of this Key, but next to the Willow leaves which it greatly resembles. In all other respects it is a true Oak, as can be seen in the Master Pages.

SCALE
½ ACTUAL SIZE

White Oak 52

Burr Oak 52

The lobes
are sometimes longer
and less rounded.

Chestnut Oak 52

Swamp White Oak 52

Red (or Black) Oak Group (bristle-pointed lobes)

SCALE
½ ACTUAL SIZE

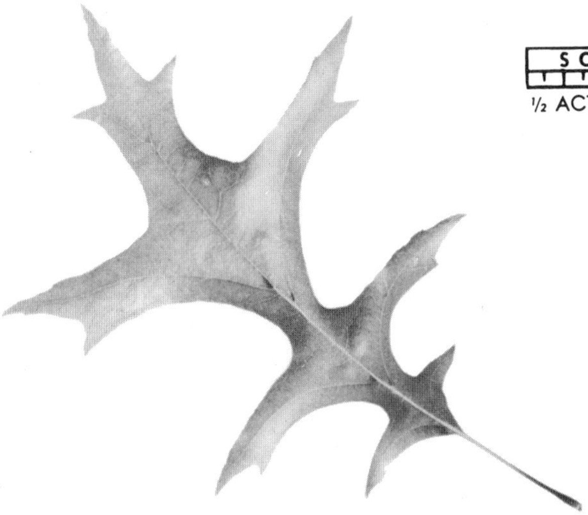

Pin Oak 52

Scarlet Oak 52

Black Oak 52

Red Oak 52

SCALE
½ ACTUAL SIZE

Red Maple 96

Sugar Maple 96

Silver Maple 96

Striped Maple 96

Sweet Gum 71

SCALE
½ ACTUAL SIZE

Norway Maple 96

Sycamore Maple 96

Mountain Maple 96

Sycamore 72

Tulip Tree 68

SCALE
ACTUAL SIZE

Redbud 80

Balsam Poplar 28

Cottonwood 28
(Poplar)

American Linden 95

White
Mulberry 60

Hawthorn 73

White Poplar 28

Miscellaneous Simple Leaves

Note: Most Poplars have flat stems.

SCALE
ACTUAL SIZE

Balm of Gilead 28
(Poplar)

Lombardy Poplar 28

Trembling Aspen 28
(Poplar)

Red Mulberry 60

sometimes very large, up to ten inches

Paper Birch 44

Large-toothed Poplar 28

Red Birch 44

Gray Birch 44

Miscellaneous Simple Leaves

SCALE
ACTUAL SIZE

Flowering Dogwood 112

Alternate-leaved Dogwood 112

American Beech 48

Nannyberry 127

American Elm 42

Miscellaneous Simple Leaves

SCALE
ACTUAL SIZE

Osage Orange 62

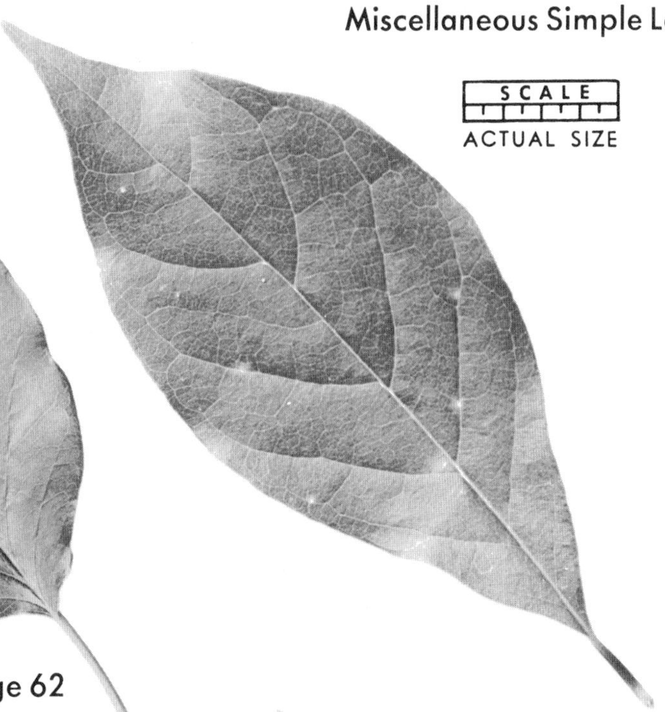

Tupelo 110

Choke Cherry 76

Black Cherry 76

Shadbush 79

Slippery Elm 42

Note: Upper side of Slippery Elm feels rough in *all* directions. Smaller teeth than American Elm.

Miscellaneous Simple Leaves

SCALE
ACTUAL SIZE

Persimmon 115

Sourwood 114

Hornbeam 50

Hackberry 41

Yellow Birch 44

Black Birch 44

Hop Hornbeam 51

Miscellaneous Simple Leaves

SCALE
ACTUAL SIZE

Willow Oak 52

Pin Cherry 76

American Chestnut 49

Black Willow 25

Weeping Willow 25

Sweet Cherry 76

Crack Willow 25

Sweet Bay 64
(Magnolia virginiana)

Ginkgo 24

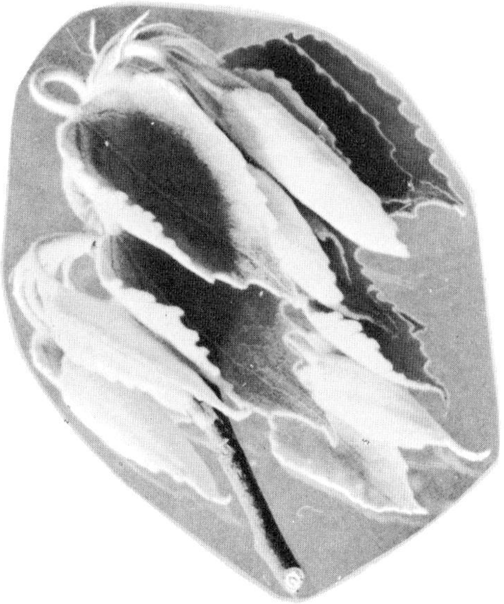

Large-toothed
Poplar 28
(immature leaves very
silvery at this stage)

Holly 94

SCALE
½ ACTUAL SIZE

Sassafras 70

SCALE
ACTUAL SIZE

White Mulberry 60

Note: These shapes are typical of Mulberries.

SCALE
½ ACTUAL SIZE

Catalpa 124

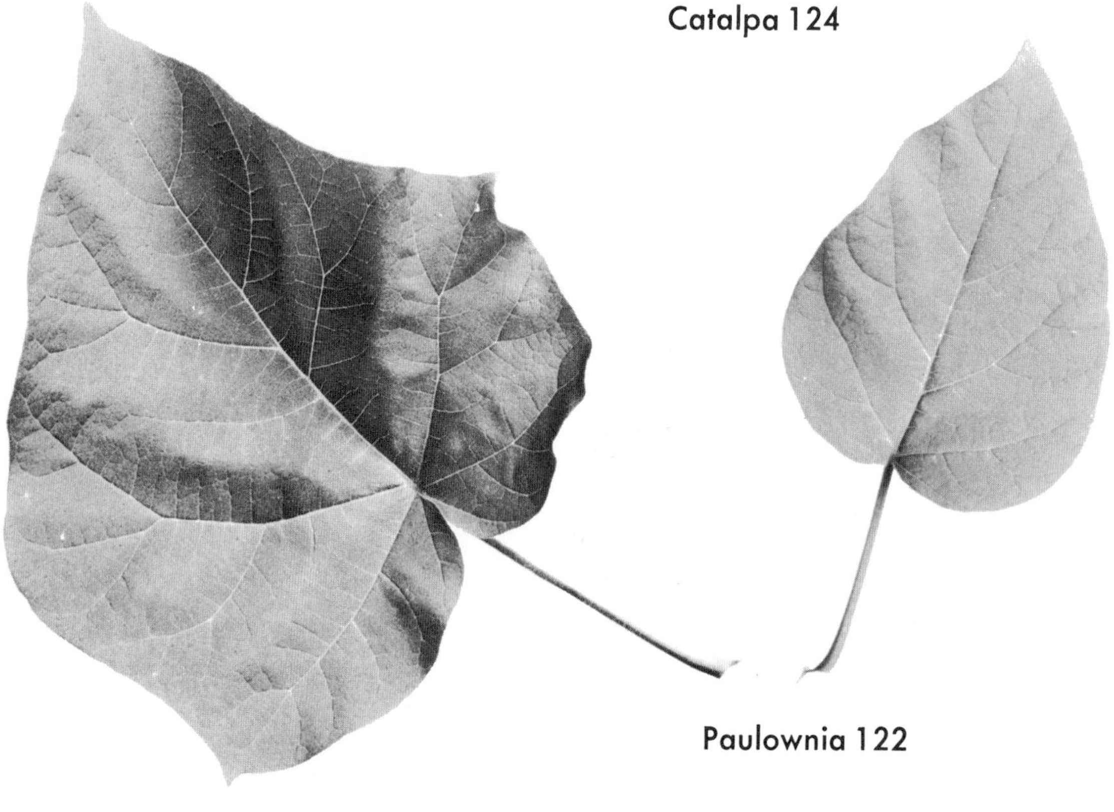

Paulownia 122

SCALE
½ ACTUAL SIZE

Umbrella Tree 64
(Magnolia tripetala)

Cucumber Tree 64
(Magnolia acuminata)

distinctive smell when crushed, as compared with relatively odorless Magnolias above

Pawpaw 63

Sassafras 70

SCALE

⅓ ACTUAL SIZE

usually
seven leaflets

Horsechestnut 106

usually
five leaflets

Buckeye 106

Pecan 34

SCALE
½ ACTUAL SIZE

Hickory 34

Hickory 34

Pinnately Compound Leaves

SCALE
½ ACTUAL SIZE

Black Walnut 38

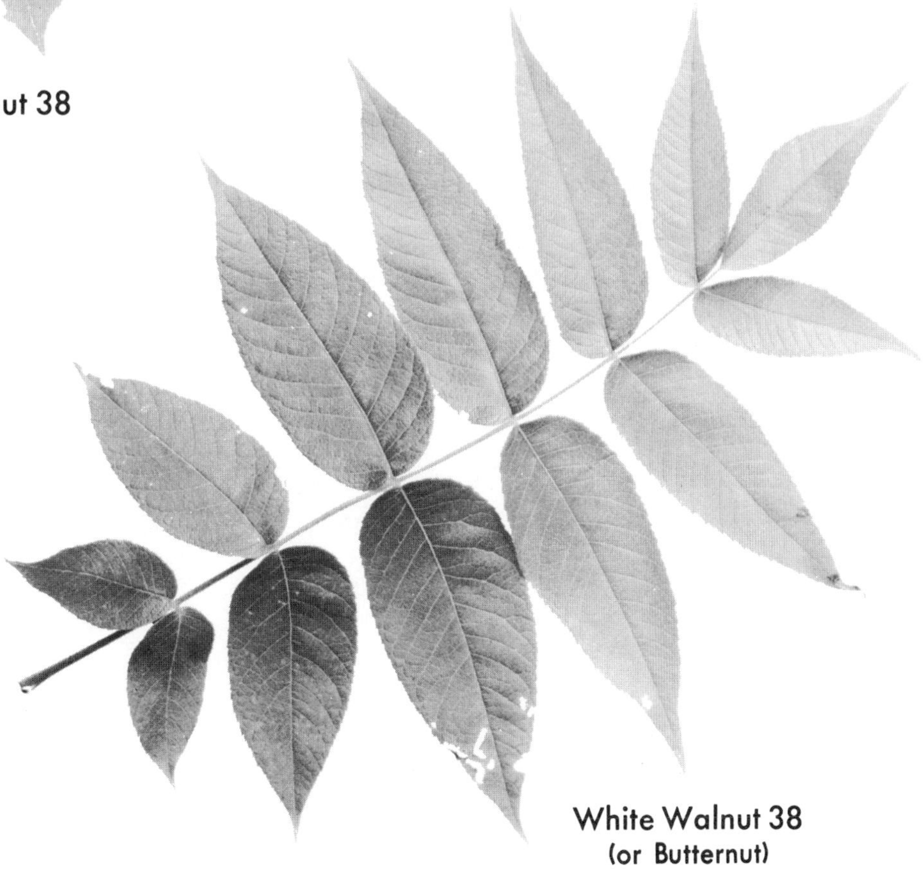

White Walnut 38
(or Butternut)

SCALE
½ ACTUAL SIZE

Black Ash 116

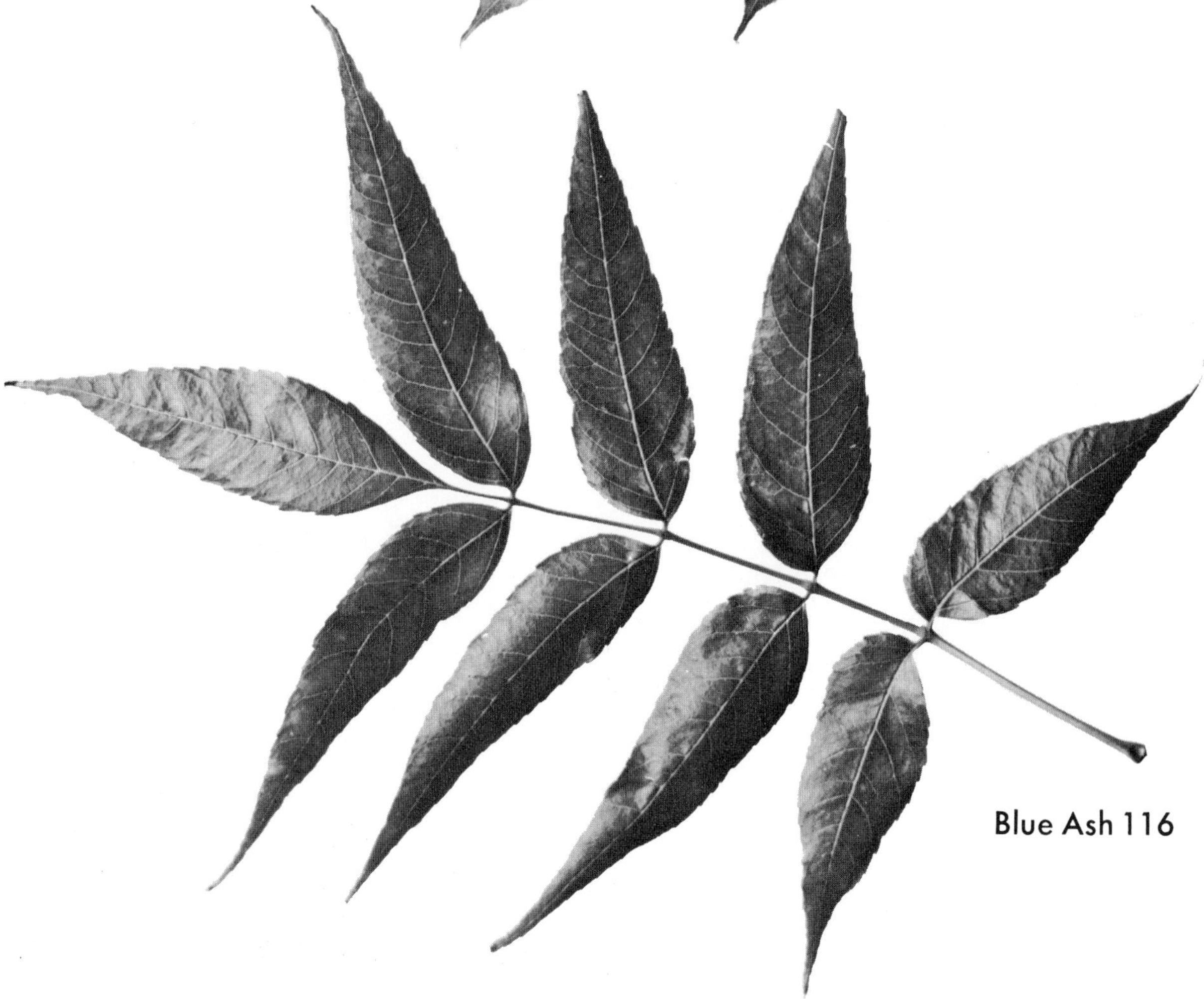

Blue Ash 116

SCALE
½ ACTUAL SIZE

Green Ash 116

Red Ash 116

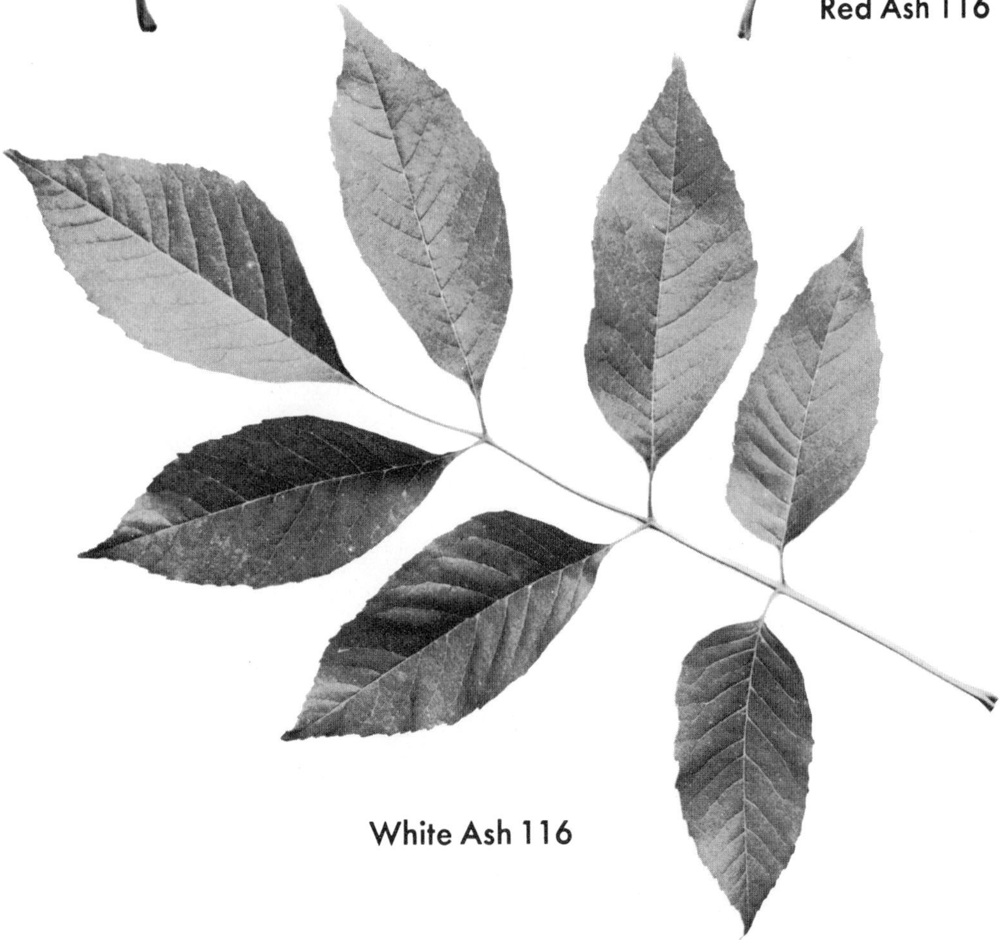

White Ash 116

Pinnately Compound Leaves

SCALE
½ ACTUAL SIZE

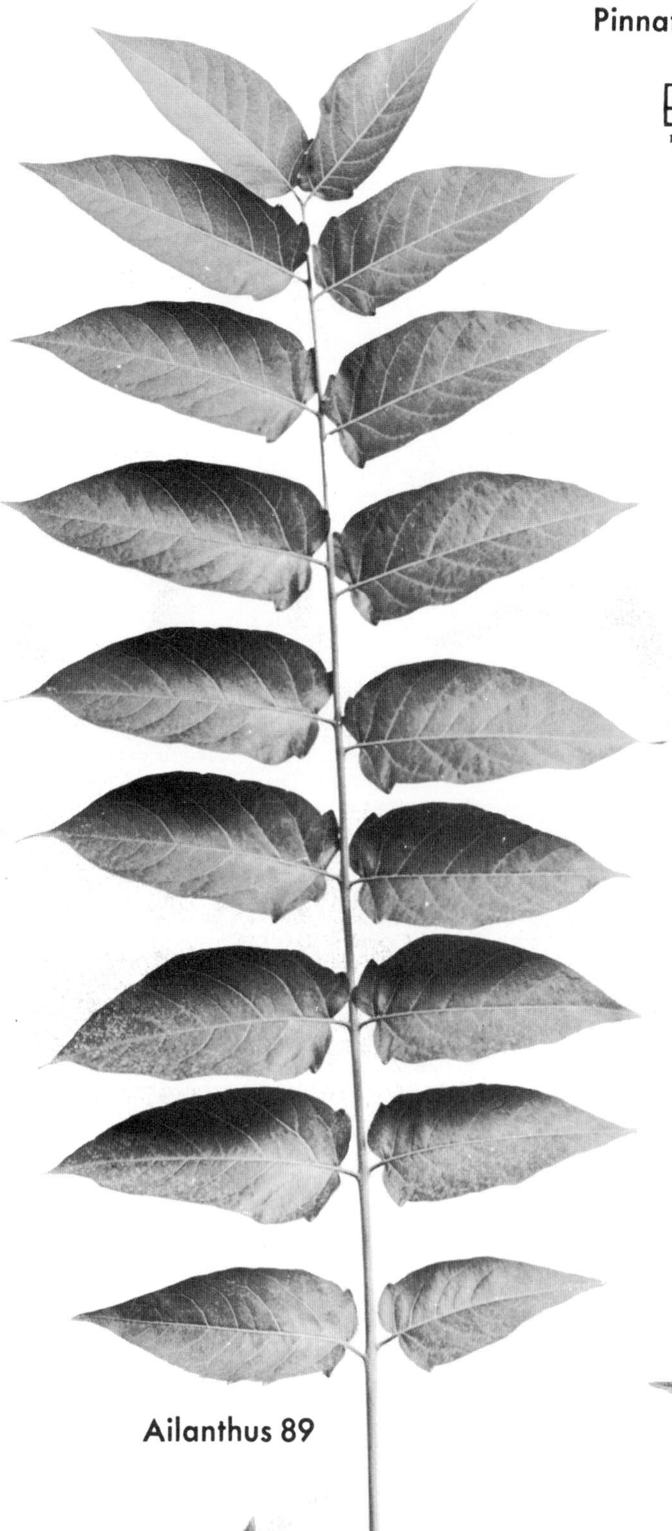

Ailanthus 89

Poison Sumac 90

Ash-leaved Maple 96
(or Box Elder)

Hop Tree 88
(or Wafer Ash)

Pinnately Compound Leaves

SCALE
½ ACTUAL SIZE

Staghorn Sumac 90
(Smooth Sumac similar)

American Mountain Ash 74

Notice wings
along stem of leaf

Dwarf Sumac 90

European Mountain Ash 74

SCALE
½ ACTUAL SIZE

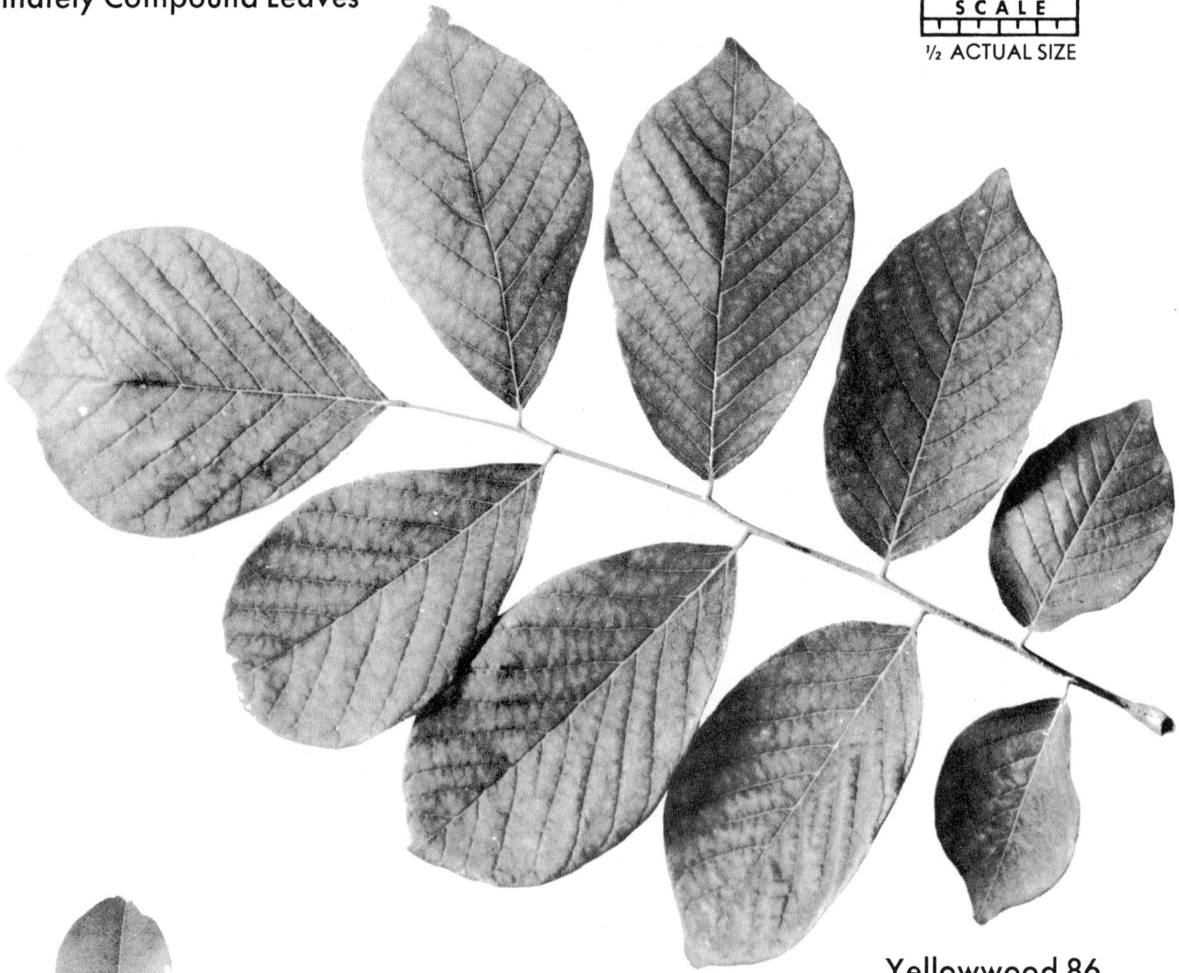

Yellowwood 86

Common (or Black)
Locust 81

Honey Locust 82
Doubly compound leaf above. The one on left is more typical, however.

Kentucky Coffee Tree 84
(doubly compound leaf is typical)

Key #4 **FLOWERS**

This Key is arranged primarily according to the sequence of bloom. The blooming season starts in March and ends in August for the trees included here. No specific dates are given, as the range may make a difference of a month or more, and even in the same locality the time of bloom is not the same every year. The sequence, however, remains approximately the same, regardless of place or year.

In a few instances, in order to place similar flowers together, the sequence has been changed slightly, but note of this has been made in each instance.

This Key is an important one, for although the blooming period of trees is short, all mature trees blossom, whereas not all trees bear fruit. Some are dioecious, which means that male (staminate) flowers are born on one tree and female (pistillate) flowers on another. Only those having female flowers will bear fruit, of course. Some trees are monoecious, bearing both male and female flowers on the same tree, while others produce *perfect* flowers, these having both male and female parts in the same flower. The last two types bear fruit on all mature trees.

Flower colors are noted where distinctive.

SCALE
ACTUAL SIZE

FL 1

red
female
(pistillate)

yellowish red
male
(staminate)

Silver Maple 96

red
female
(pistillate)

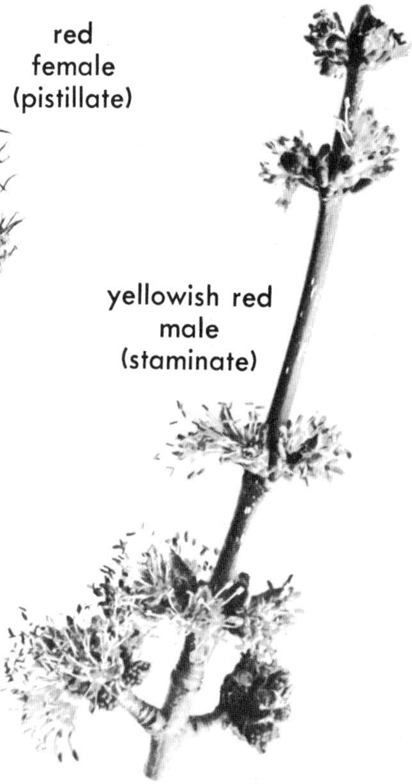

yellowish red
male
(staminate)

Red Maple 96

Elm 42

Cottonwood 28
female (pistillate)

SCALE
ACTUAL SIZE

Hop Hornbeam 51

Poplar 28

Hornbeam 50

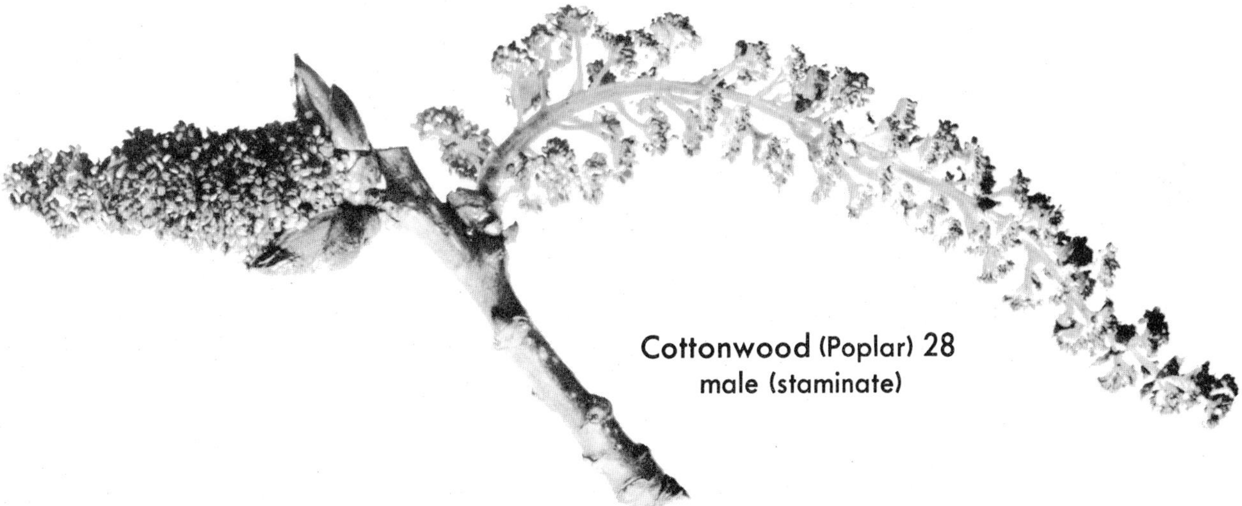

Cottonwood (Poplar) 28
male (staminate)

SCALE
ACTUAL SIZE

FL
3

Willow 25

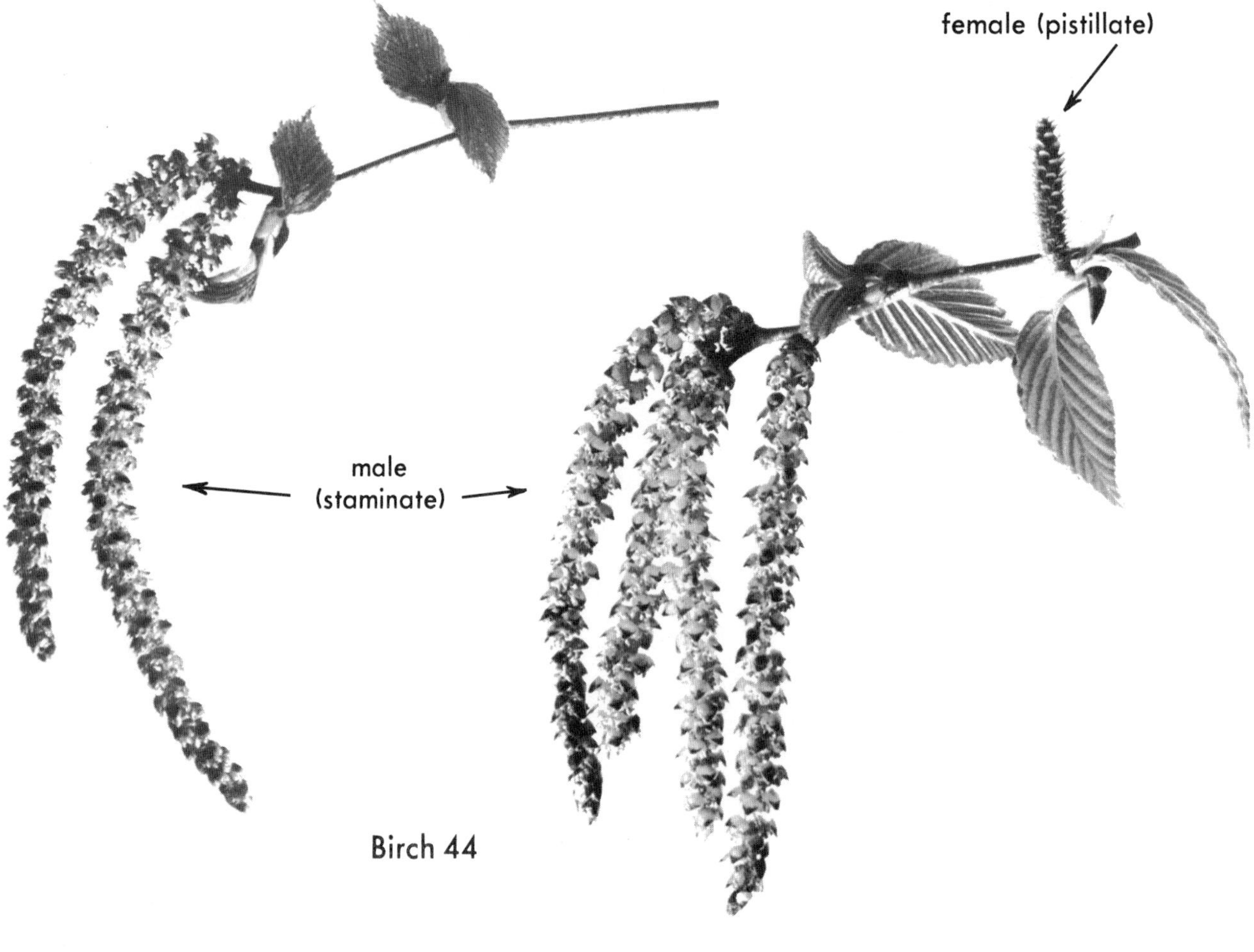

Birch 44

SCALE
ACTUAL SIZE

Sweet Cherry 76

Ginkgo 24

Pin Cherry 76

Redbud 80

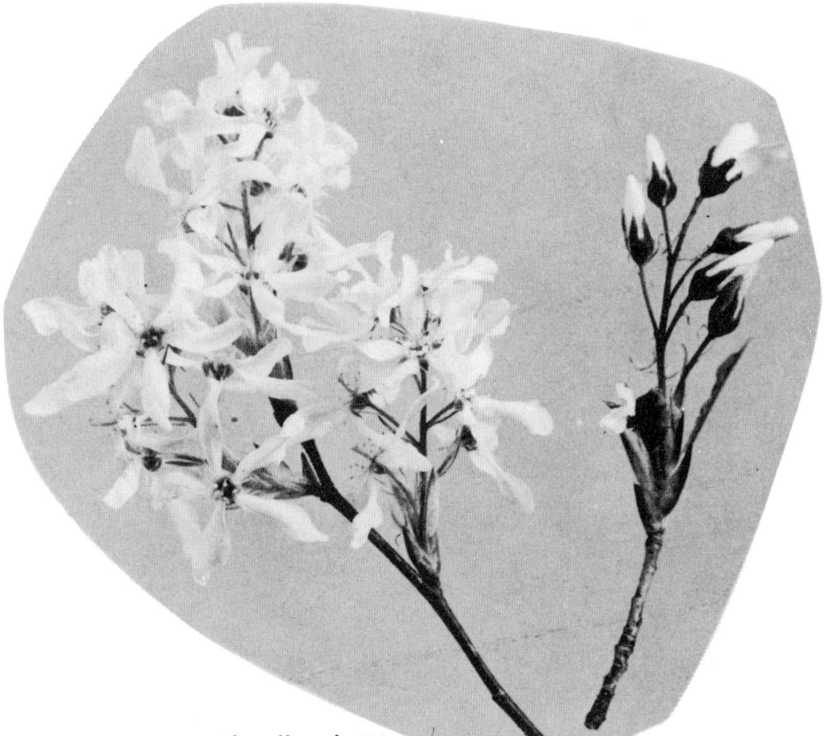

Shadbush 79

SCALE
ACTUAL SIZE

red

male
(staminate)

yellowish

female
(pistillate)

Ash-leaved Maple 96
(Box Elder)

Striped Maple 96
yellow

Beech 48

Sassafras 70
(yellowish)

Sugar Maple 96

(yellow)

SCALE

ACTUAL SIZE

Norway Maple 96

(yellow)

FL 6

Flowering Dogwood 112

Hackberry 41

Sweet Gum 71

SCALE
ACTUAL SIZE

female (pistillate)

male (staminate)

Ash 116

(Most Ashes are dioecious: male and female flowers on different trees)

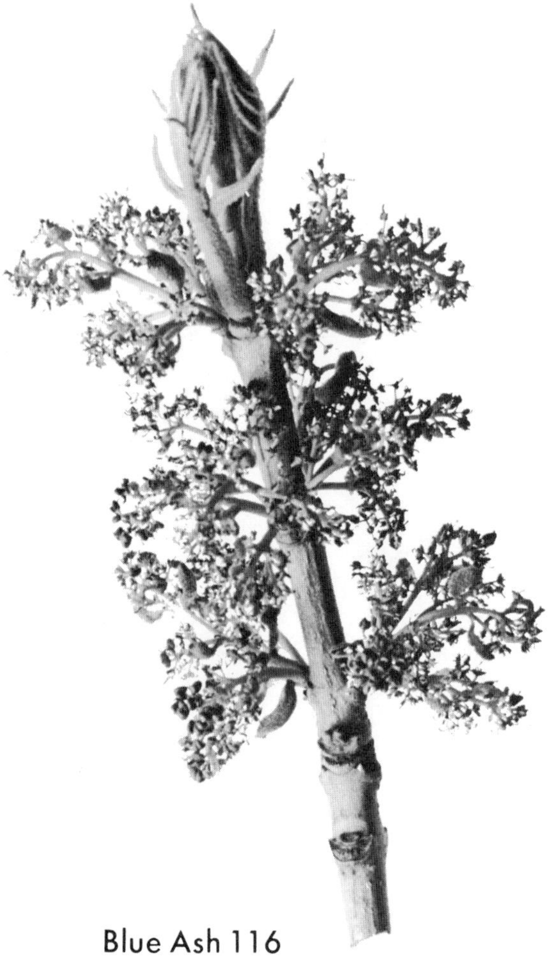

Blue Ash 116
perfect flowers (male and female parts in same flower)

Pawpaw 63

SCALE
ACTUAL SIZE

Hickory 34

Oak 52

Walnut 38

SCALE
½ ACTUAL SIZE

yellowish

Ohio Buckeye 106

lavender

Paulownia 122

yellowish

Sweet Buckeye 106

white flecked with red

Horsechestnut 106

SCALE

ACTUAL SIZE

Choke Cherry 76

Black Cherry 76

The Choke Cherry blooms about a week to ten days before the Black Cherry.

Mountain Maple 96

Note: Viburnums have five petals; Dogwoods have four petals (not readily evident in photographs).

SCALE
ACTUAL SIZE

Nannyberry 127
(Sweet Viburnum)

Alternate-leaved Dogwood 112

Hawthorn 73

European Mountain Ash 74

SCALE
ACTUAL SIZE

male (staminate)
flowers

female (pistillate)
flowers

Tupelo 110

American Mountain Ash 74

SCALE

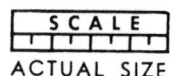

ACTUAL SIZE

FL 13

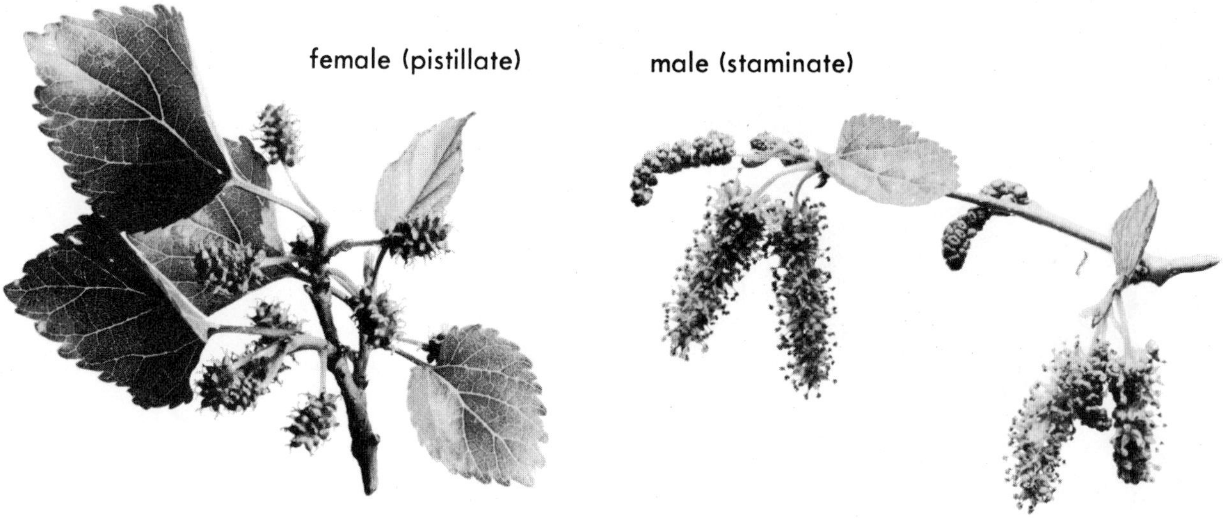

Mulberry 60

Sycamore 72

Osage Orange 62

SCALE
½ ACTUAL SIZE

Yellowwood 86

Sycamore Maple 96
yellowish

SCALE
ACTUAL SIZE

Common or Black Locust 81

SCALE

ACTUAL SIZE

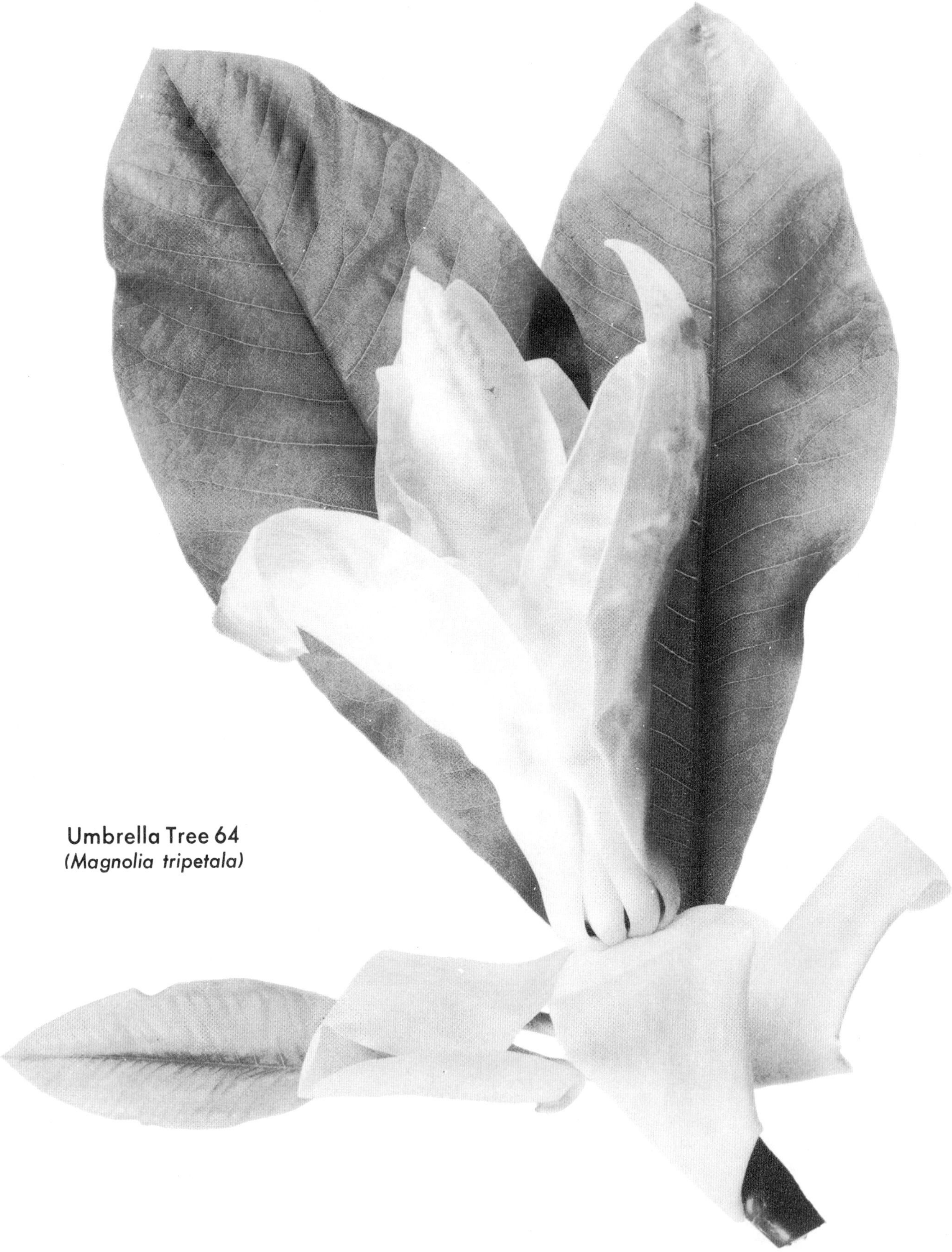

Umbrella Tree 64
(Magnolia tripetala)

(greenish petals)

Cucumber Tree 64
(Magnolia acuminata)

Sweet Bay or Swamp Magnolia 64
(Magnolia virginiana)

SCALE
ACTUAL SIZE

FL
17

Tulip Tree 68

Hop Tree 88

SCALE

ACTUAL SIZE

American Holly 94

whitish
male (staminate)

light-yellow

Persimmon 115

female (pistillate)

Kentucky Coffee Tree 84

SCALE
½ ACTUAL SIZE

Ailanthus 89

Honey Locust 82

SCALE
ACTUAL SIZE

Poison Sumac 90

Catalpa 124

SCALE
ACTUAL SIZE

Staghorn Sumac 90

SCALE
ACTUAL SIZE

Staghorn Sumac 90

Japanese Dogwood 112

American Chestnut 49

SCALE
ACTUAL SIZE

American Linden 95

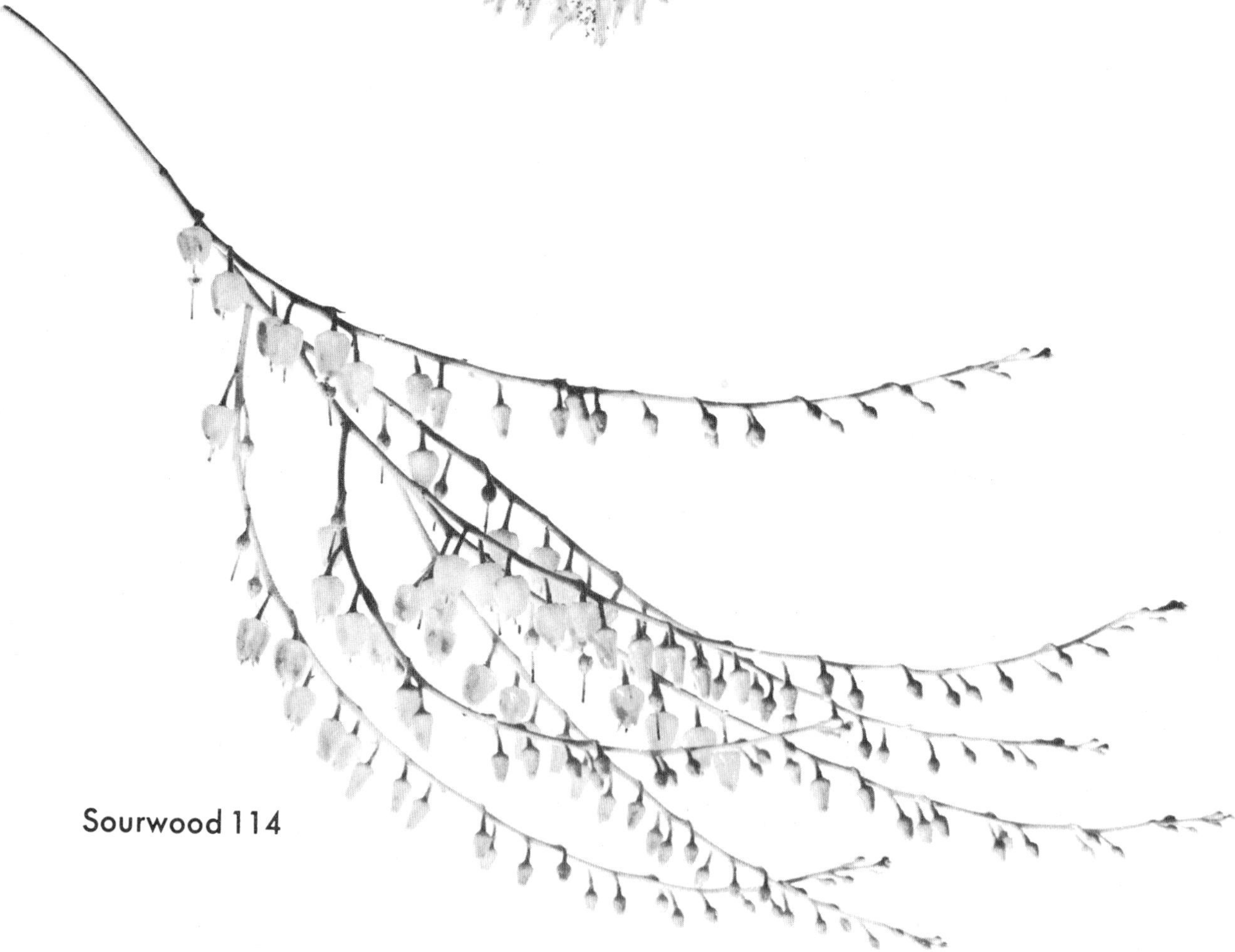
Sourwood 114

SCALE
2/3 ACTUAL SIZE

Smooth Sumac 90

red

red

Dwarf Sumac 90

Key #5 **FRUIT**—All actual size

Fruit is the seed-bearing product of a plant

This Key is arranged to compare similar-appearing fruit. The headings, although not always technically correct, were chosen as being most descriptive of the fruit pictured.

- Nuts—showing the husk as well as the actual nut; burrs, acorns and similar-appearing fruit
- Pods
- Winged Seeds
- Berries and similar fruit
- Odd and miscellaneous fruit

Look carefully for fruit, as it often hangs on the tree for most of the year. Don't just look up, look down also, and at your feet may be the clue you couldn't reach or see clearly. It is not wise to depend on material found on the ground, as it may not belong to the tree under which you found it. However, don't overlook evidence, it can start you on the right track.

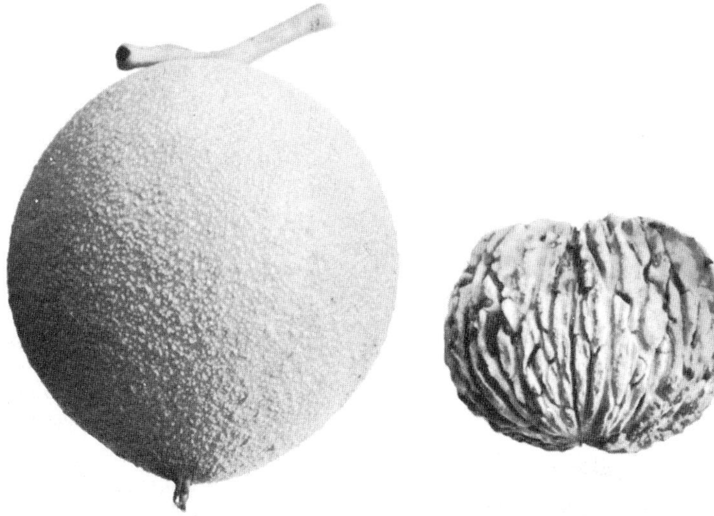

Black Walnut 38

Pecan 34

Bitternut Hickory 34

Butternut or White Walnut 38

Pignut Hickory 34

Shagbark Hickory 34

Mockernut Hickory 34

Nuts

Horsechestnut 106

Sweet Buckeye 106
(thick husk)

Ohio Buckeye 106
(thin husk)

American Chestnut 49

Beech 48

Burr Oak 52

Swamp White Oak 52

Black Oak 52

Scarlet Oak 52

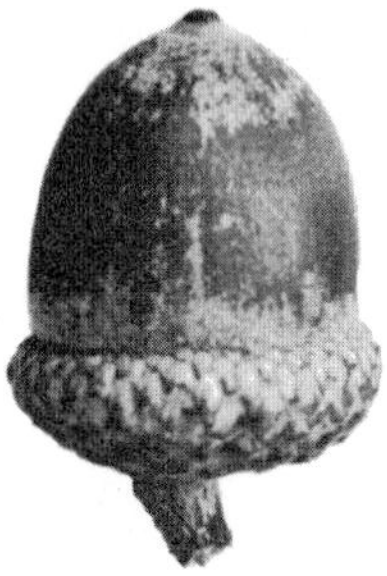

Red Oak 52

Chestnut Oak 52

White Oak 52

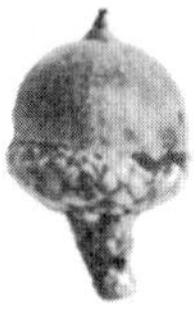

Pin Oak 52

Willow Oak 52

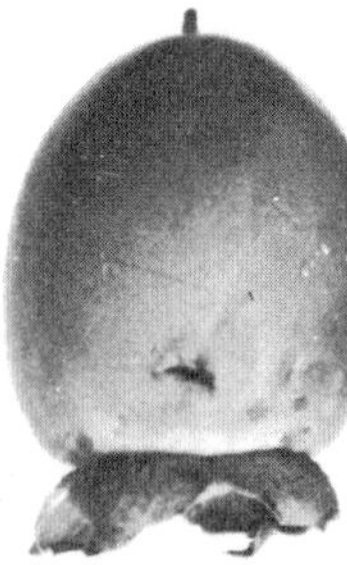

Persimmon 115

This is not an acorn. It has no cup and does not have a shell. The Persimmon is edible and juicy, but shouldn't be eaten until ripe, as it is very astringent.

winged seeds

Paulownia 122

This is not a nut, but might be mistaken for one. Actually, it is a pod, which contains many tiny winged seeds.

Catalpa 124

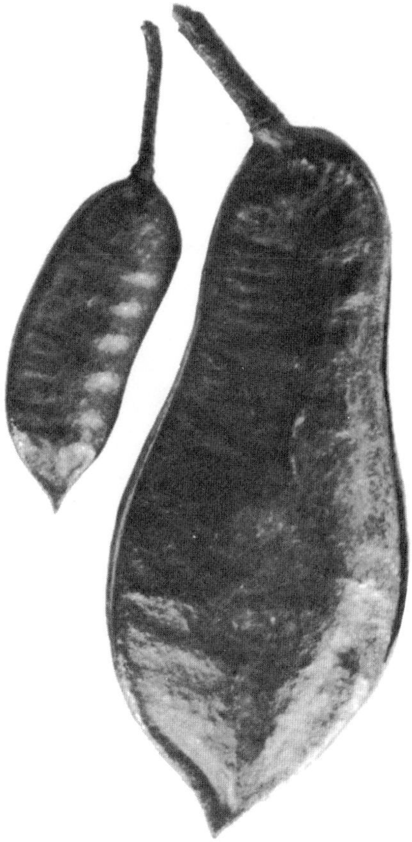

Kentucky Coffee Tree 84

Honey Locust 82

Pods

Redbud 80

Yellowwood 86

Common or Black Locust 81

Swamp Locust 82

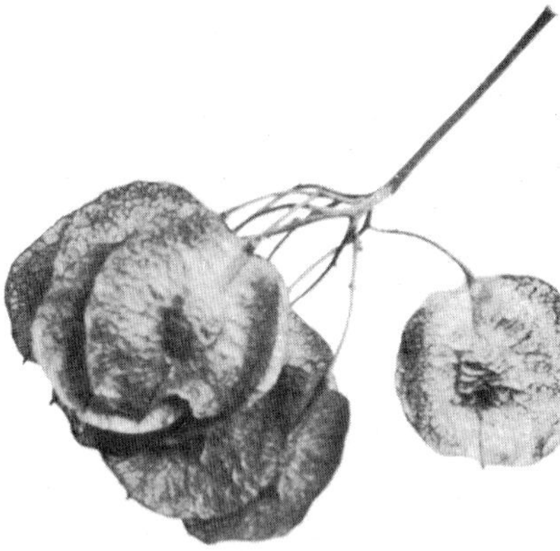

Hop Tree 88

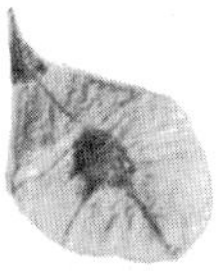

Slippery Elm 42

American Elm 42

Ailanthus 89

Hornbeam 50

Hop Hornbeam 51

Winged Seeds

Blue Black Red Green White

Ash 116

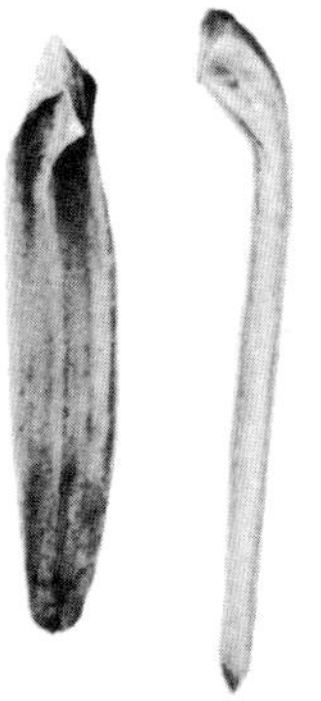

Tulip Tree 68
(see also Fruit 14)

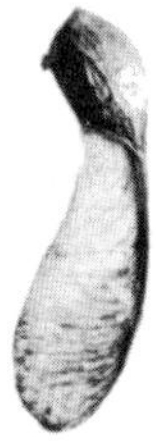

Maple 96
(see opp. page)

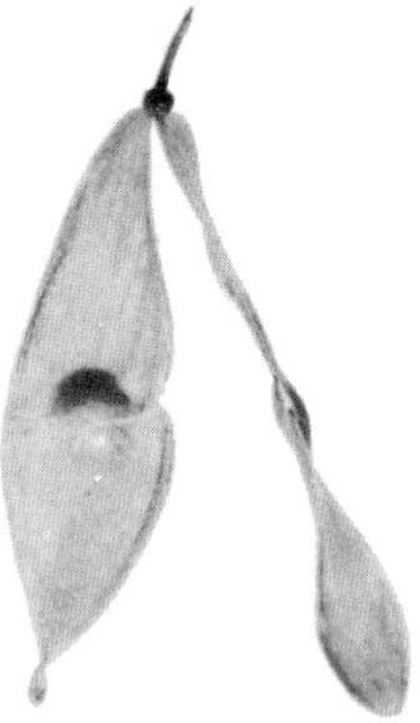

Ailanthus 89
(see also Fruit 6)

cluster of Ash seeds 116

cluster of Maple seeds 96

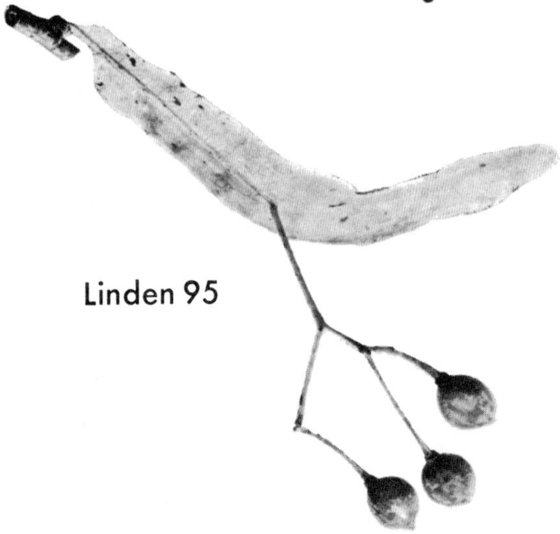

Linden 95

Catalpa 124
(winged seed from pod—see Fruit 4)

Ash-leaved Maple 96
(immature—note mature seed on opp. page)

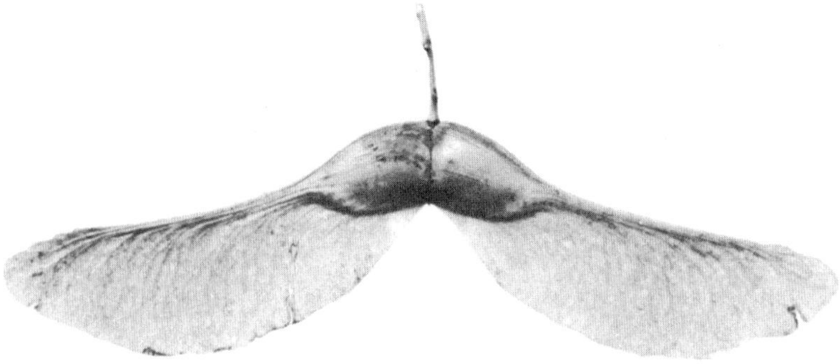

Norway Maple 96

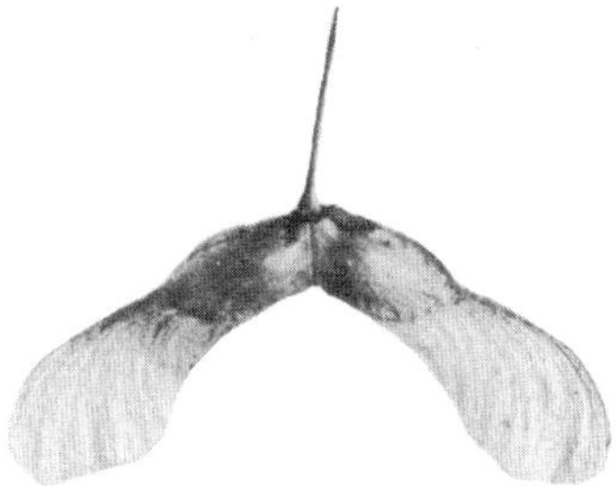

Striped Maple 96

Sycamore Maple 96

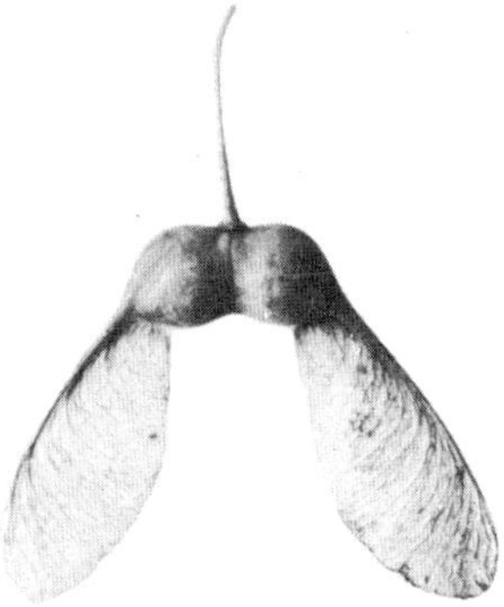

Sugar Maple 96

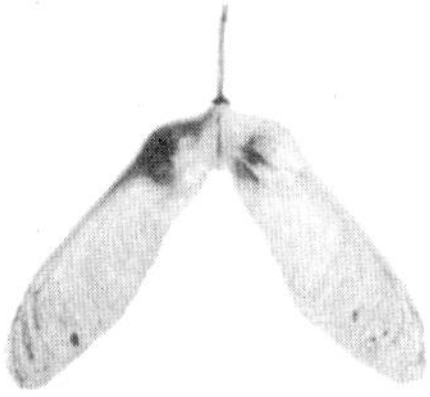

Mountain Maple 96

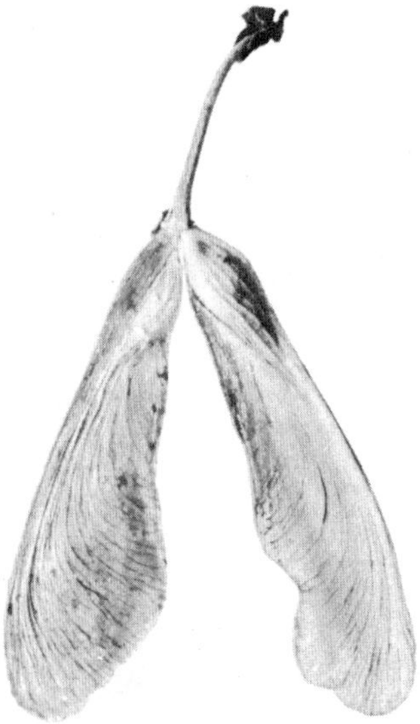

Silver Maple 96

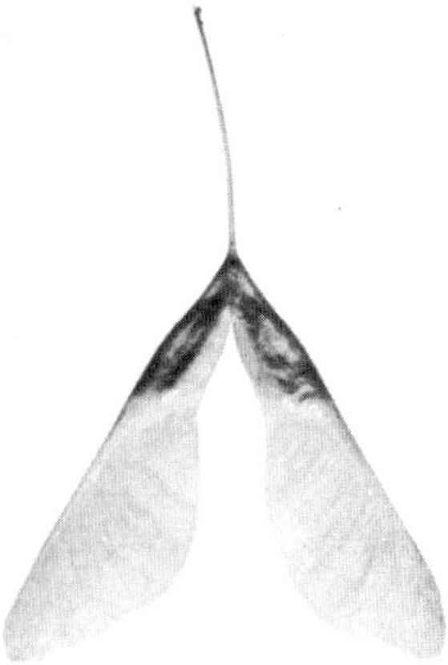

Ash-leaved Maple 96
(or Box Elder)

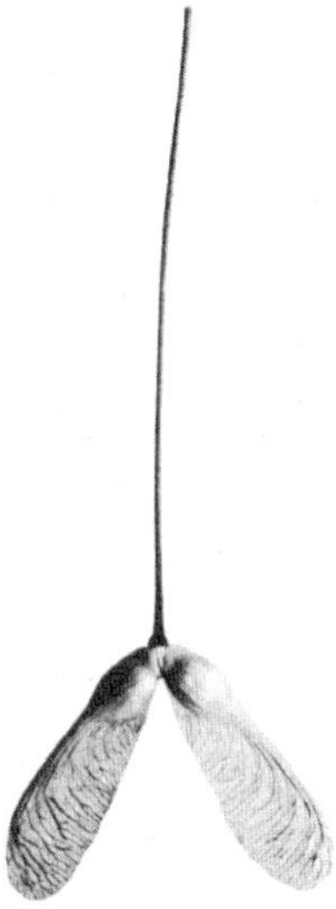

Red Maple 96

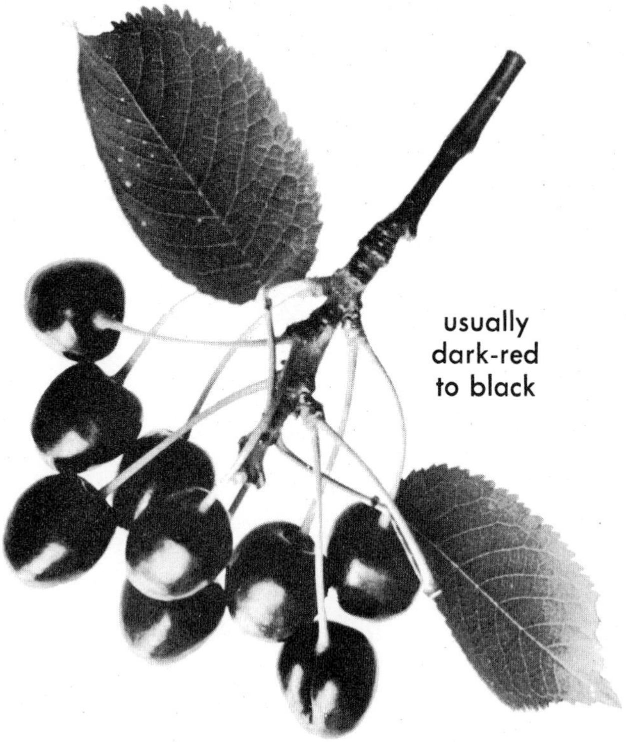

usually dark-red to black

Sweet Cherry 76

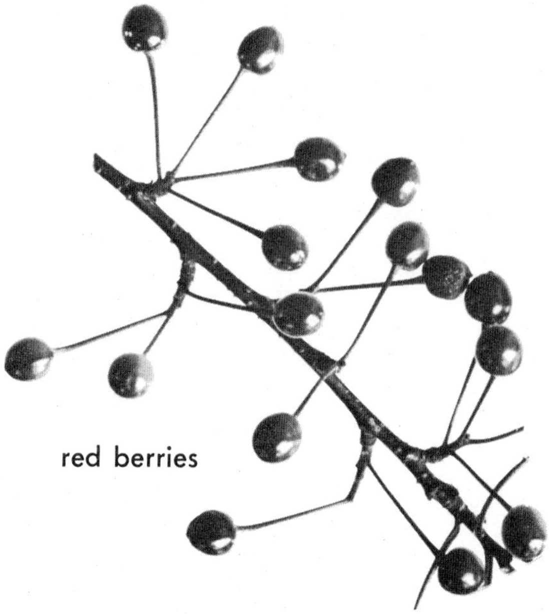

red berries

Pin Cherry 76

Choke Cherry 76

dark-red to black berries

Black Cherry 76

blue-black
berries

Tupelo 110

American Linden 95

dark-blue berries
held in bright-red
cups on red stems

Sassafras 70

red berries

Hackberry 41

red
berries

Flowering Dogwood 112

blue-black
berries

Alternate-leaved Dogwood 112

Berries

white berries: (see Master Pages for varieties or hybrids with red or black berries)

Red Mulberry 60

White Mulberry 60

Nannyberry 127

Hawthorn 73

Ginkgo 24

Persimmon 115

red

American Mountain Ash 74

Poison Sumac 90

whitish

orange-red

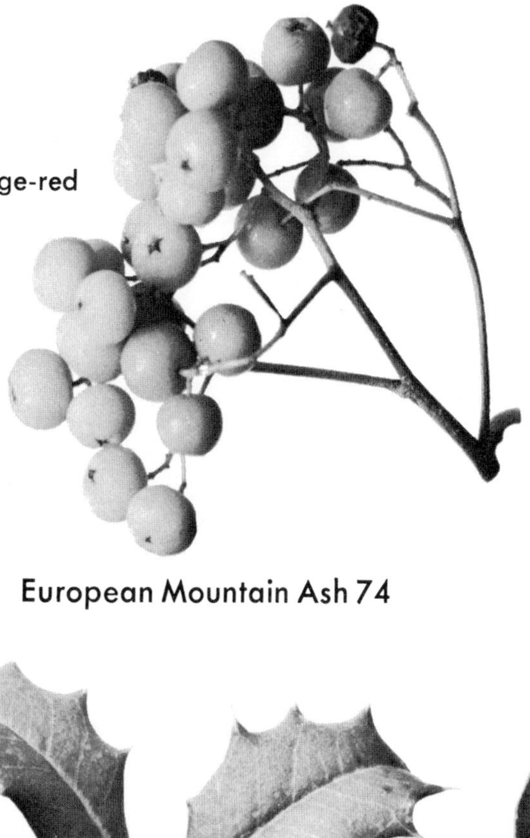

European Mountain Ash 74

pinkish-red

red

American Holly 94

Shadbush 79

red

Dwarf Sumac 90

red

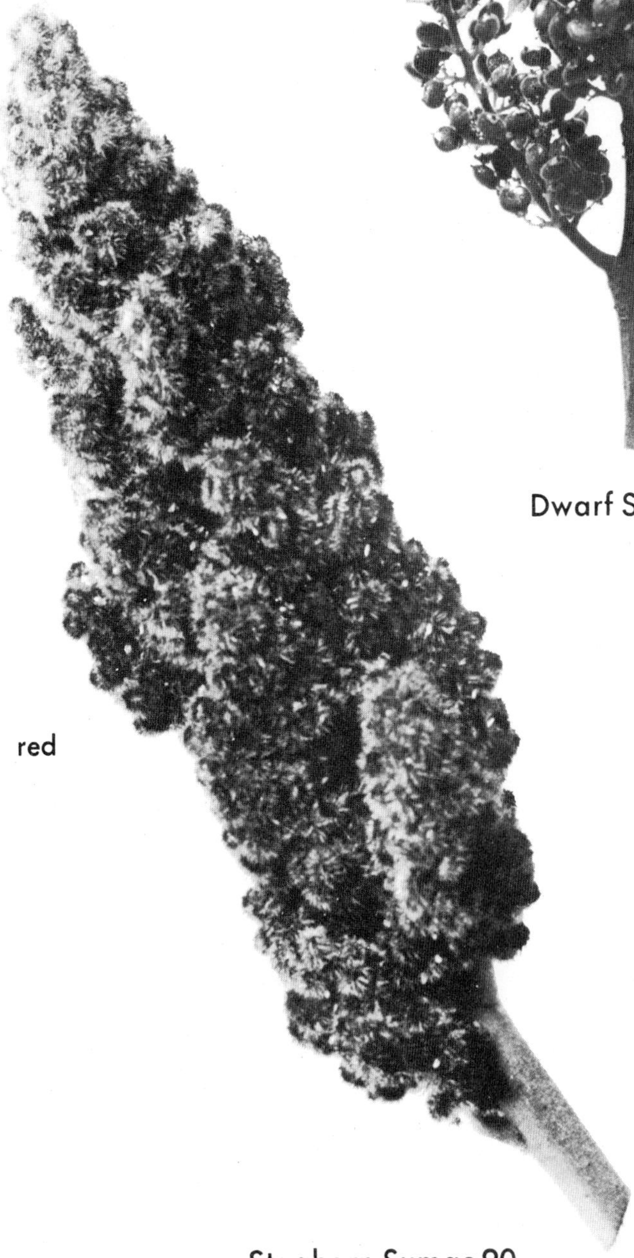

Staghorn Sumac 90

red

Smooth Sumac 90

Sweet Bay 64
(Magnolia virginiana)

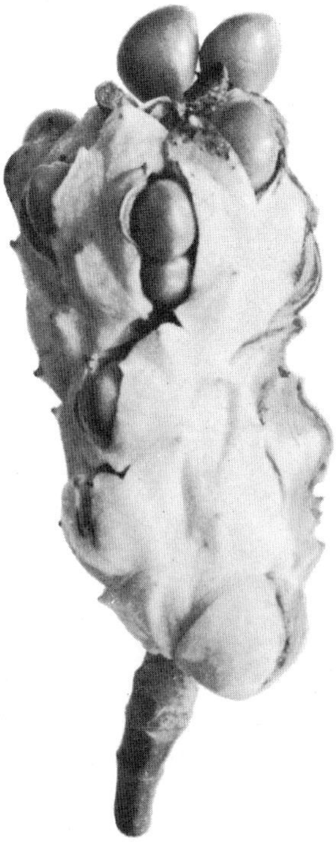

Cucumber Tree 64
(Magnolia acuminata)

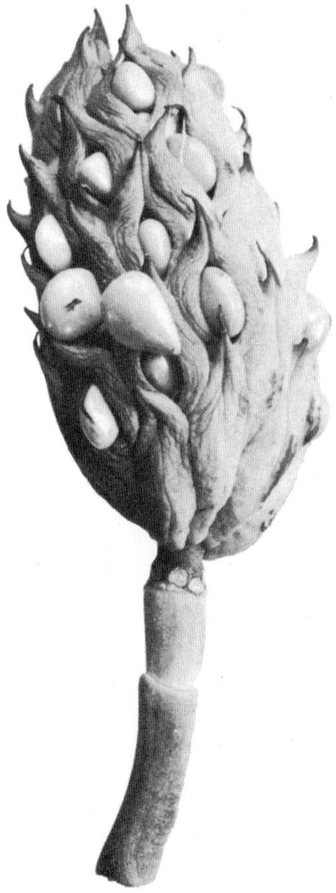

Umbrella Tree 64
(Magnolia tripetala)

Tulip Tree 68

Sycamore 72

Sweet Gum 71

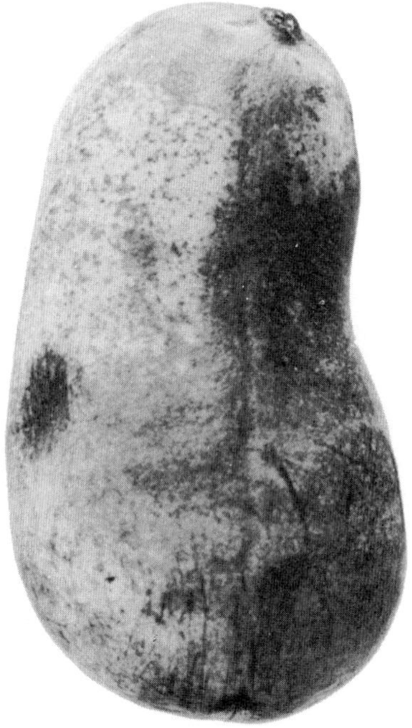

Pawpaw 63

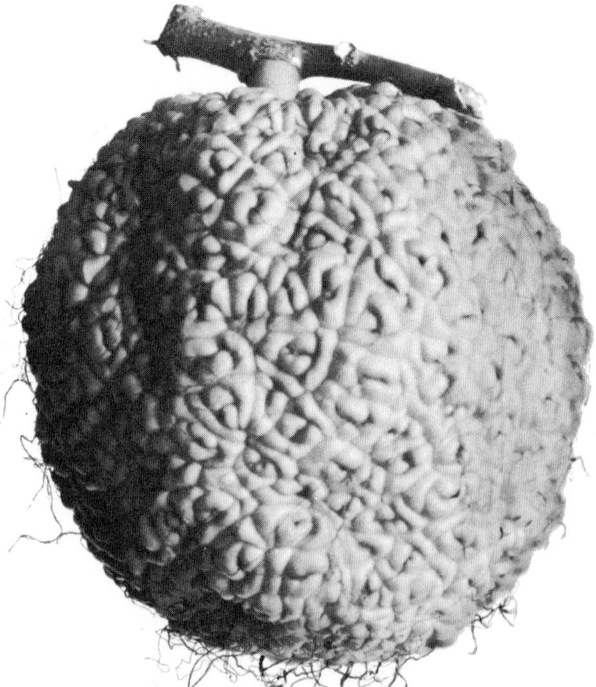

Osage Orange 62

Hop Hornbeam 51

Sourwood 114

Hornbeam 50

Cottonwood (Poplar) 28

notice cotton which blows away carrying the seed

Willow 25

Large-toothed Poplar 28
(typical of Poplars)

notice cotton which blows away carrying the seed

Typical fruit
Black, Red and Yellow Birches
44

This is *not* a fruit, but actually a male (staminate) Ash flower malformed by insects or disease. It has been placed here because, like some fruits, it often hangs on all winter and is most distinctive then. See Ash 116.

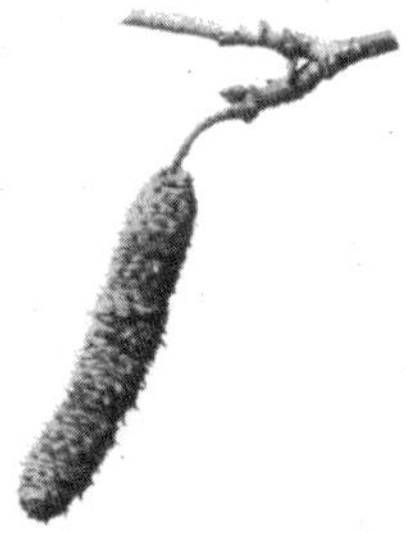

Gray Birch 44

Paper Birch 44

Key #6 **TWIGS AND BUDS**—All actual size

Twigs and Buds are the most helpful features to look for in the winter, and as an aid to their proper identification the following details are especially important to note.

Buds

- Opposite (opposite each other along twig—see Key #1)
- Size, shape, color (Note that flower buds are usually larger than growth buds and in some cases are very distinctive, as those of the Flowering Dogwood which alone identify the tree. They are not always present, however, so learn to recognize both kinds.)
- End buds (single or many-clustered)
- Lateral buds (those along twig)
  - Sticking out (divergent)
  - Close to stem (appressed)

Twigs

- Opposite (small twigs opposite along main twig—see Key #1)
- Size, shape, color
- Zigzag or straight
- Smooth or hairy
- General effect: many small or a few large twigs
- Objects attached: fruit (Key #5); thorns (Key #2)
- Special features:
  - Leaf scar (place where old leaf fell off): notice size and shape, whether opposite or alternate
  - Pith: color, chambered (divided into definite sections easily seen by slitting a twig with a knife)
  - Smell, taste, milky or other type of juices

New and Old Growth (one- and two-year-old wood)

To find growth of previous season, start at the end of twig and follow back to point where bark of twig changes color and texture. There is usually a ring around the twig at this point. From this point out to the end of the twig is the growth of the latest season. This part, taken in winter, is the part usually shown in this Key. The latest growth (one-year wood) usually has only buds along it, whereas the two-year wood usually has twigs branching off from it. See Tupelo twig on opposite page.

In selecting twigs pick healthy ones. All the pictures in this Key are actual size, but this merely indicates an average healthy twig. Look at several twigs before selecting one or more. Knick the bark: a live twig, even in winter, will be green under the bark. Usually those at the outer ends of branches that get light and air are the best.

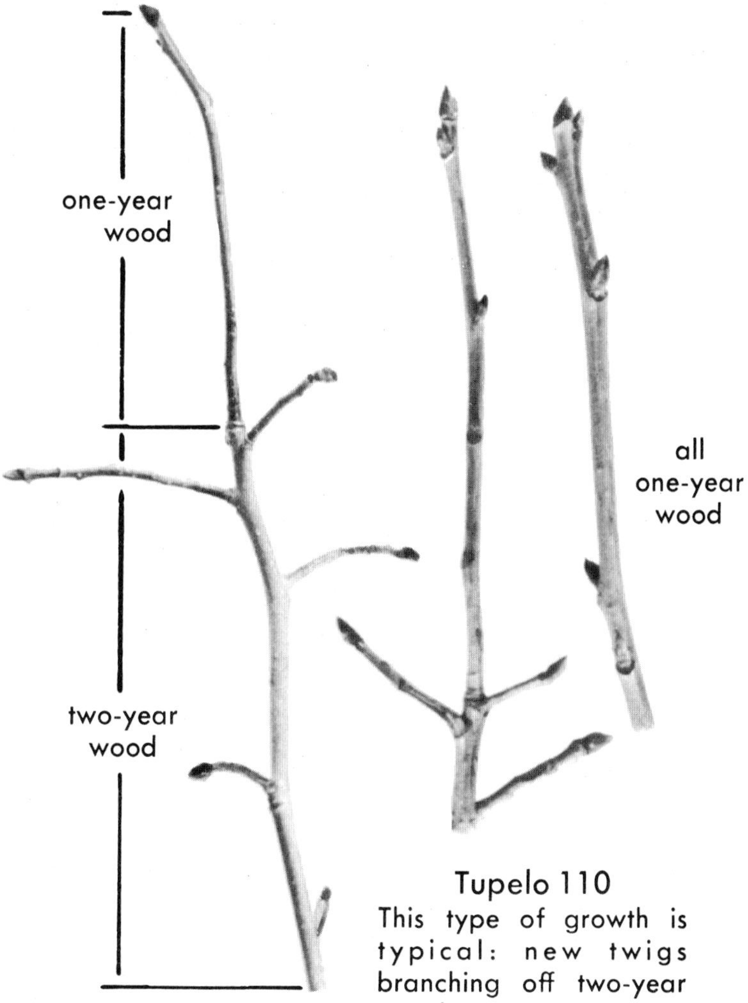

Tupelo 110

This type of growth is typical: new twigs branching off two-year wood.

Sassafras 70

Smooth, olive-green, aromatic twig. This is the exception to normal growth characteristics: side twigs often grow from the new (one-year) wood.

Cherry 76

Oak 52

Poplar 28

These trees often have a cluster of buds at the ends of the twigs. Very typical of Oaks. Cherry twigs have a disagreeable odor. Poplar buds are often very sticky.

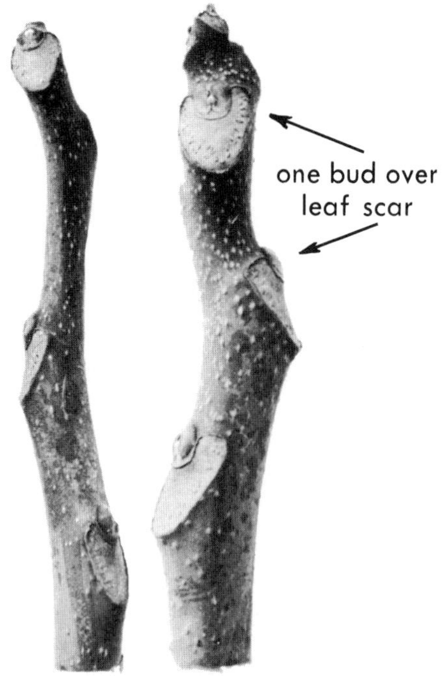

Ailanthus 89

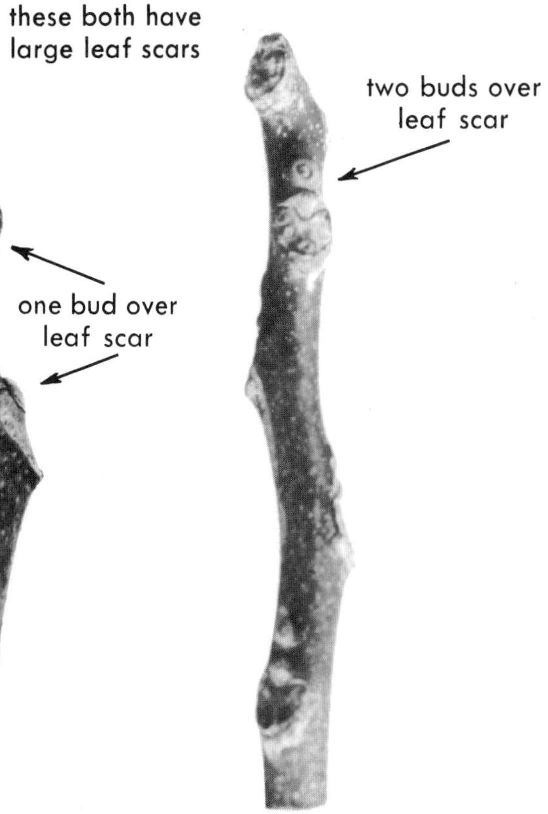

Kentucky Coffee Tree 84

Sumac 90

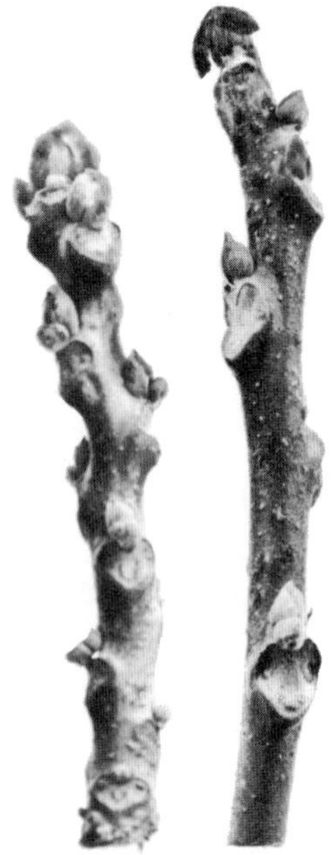

Black Walnut 38

Butternut 38
(or White Walnut)

Hickory 34

OPPOSITE—Growth (see Key #1)

Notice opposite buds, apparent on all but Dogwood, Catalpa and Paulownia

buds sticky

buds not sticky

Horsechestnut 106

Buckeyes 106

Ash 116

Maple 96

whorls of three leaf scars (the third one on other side of twig)

opposite leaf scars

only dogwood with alternate leaves and branching

flower buds

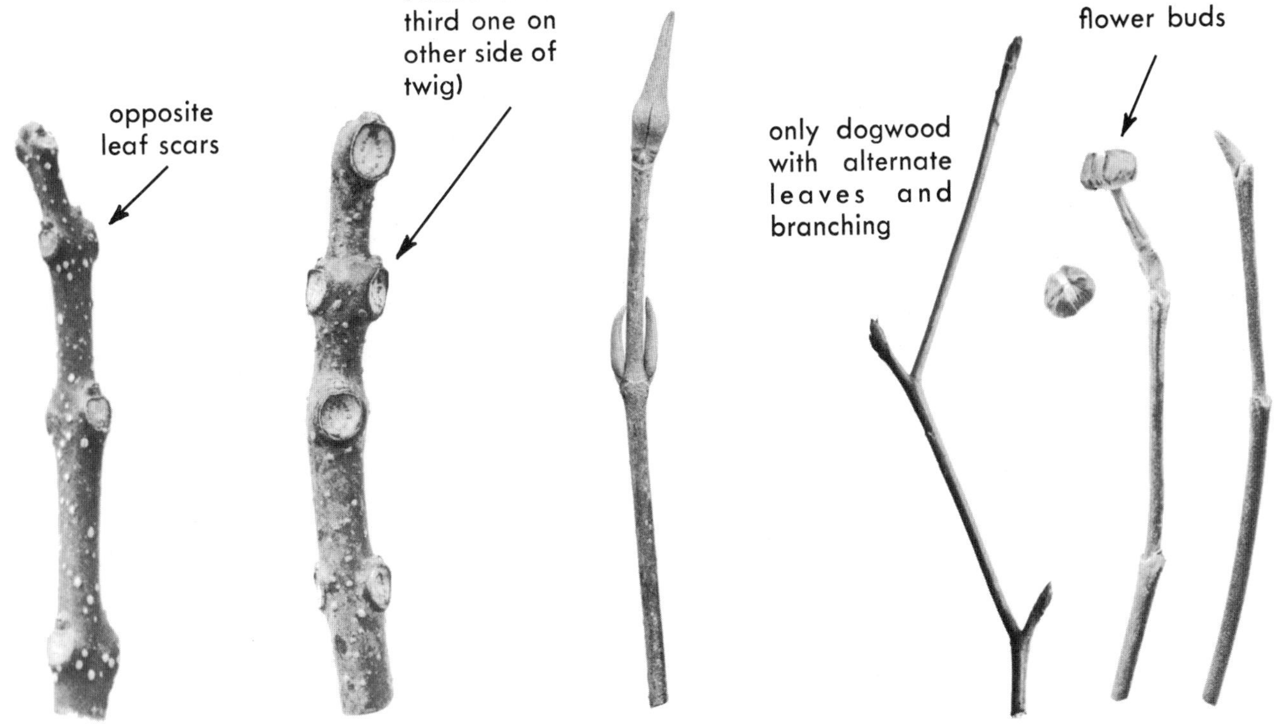

Paulownia 122

Catalpa 124

Nannyberry 127

Alternate-leaved Dogwood

Flowering Dogwood

Dogwood 112

sharp, long, narrow buds

sharp but thicker buds

buds stick
out from
stem
(divergent)

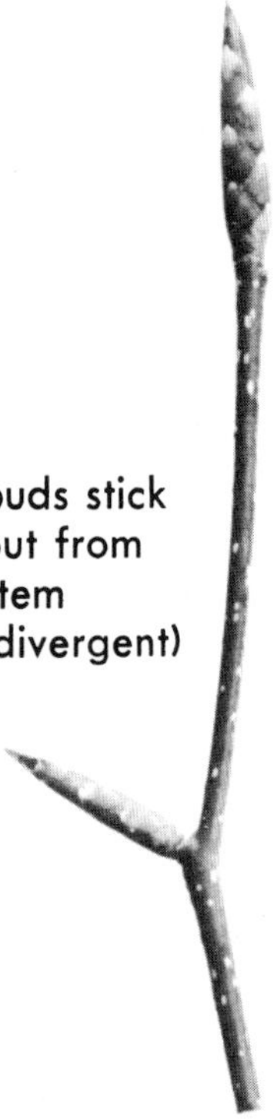

Beech 48

buds close
to stem
(appressed)

Shadbush 79

buds
sometimes
grow on
short stems
which makes
them look
divergent

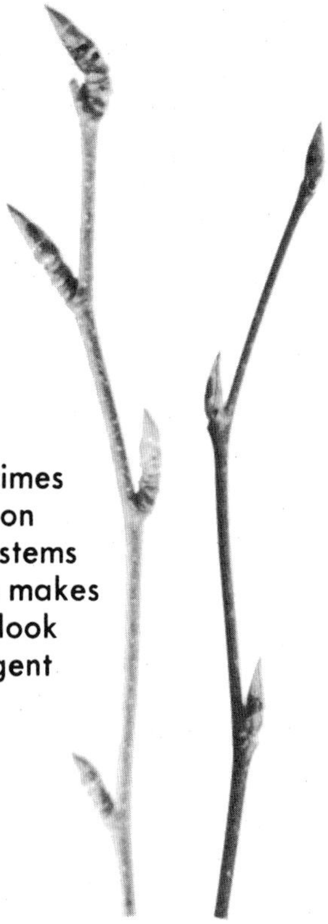

Birch 44

TW
4

note catkins

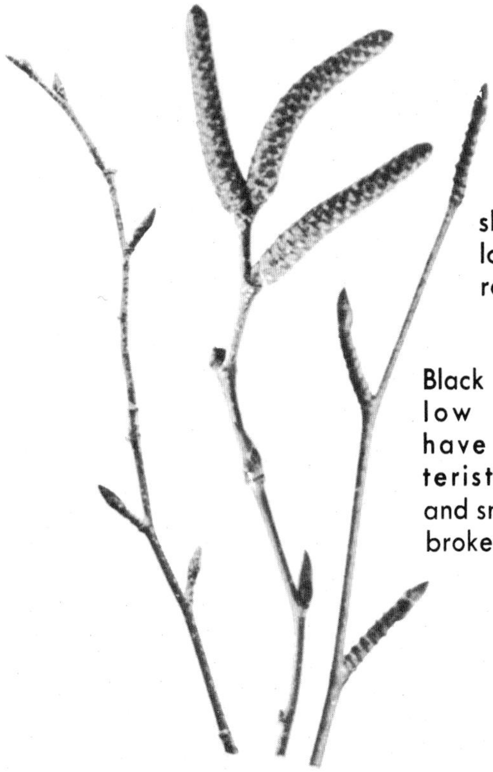

sharp but *not* long and narrow buds

Black and Yellow Birches have characteristic taste and smell when broken.

Birch 44

shiny buds

Poplar 28

Black and Choke Cherry twigs have very unpleasant taste and smell when broken.

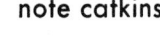

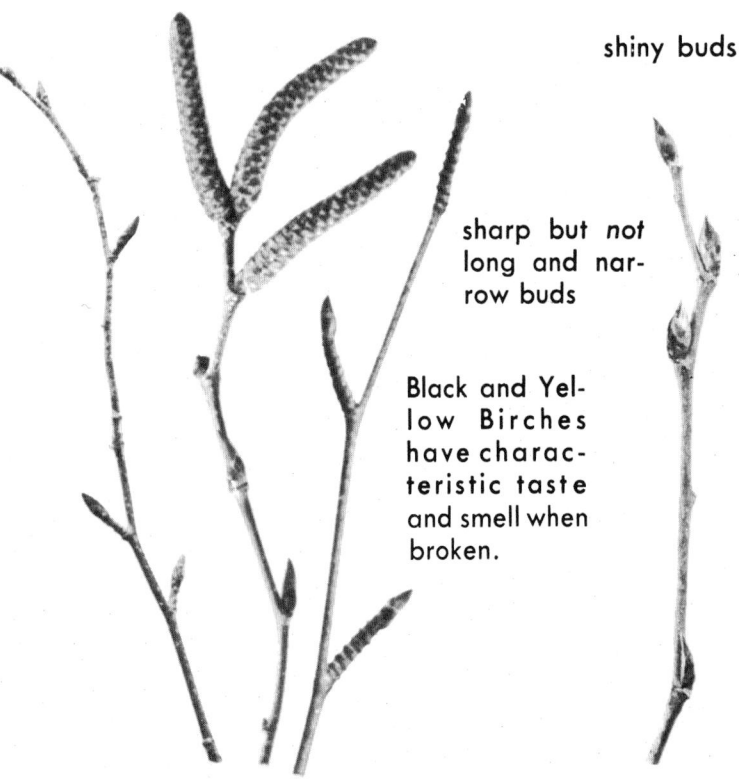

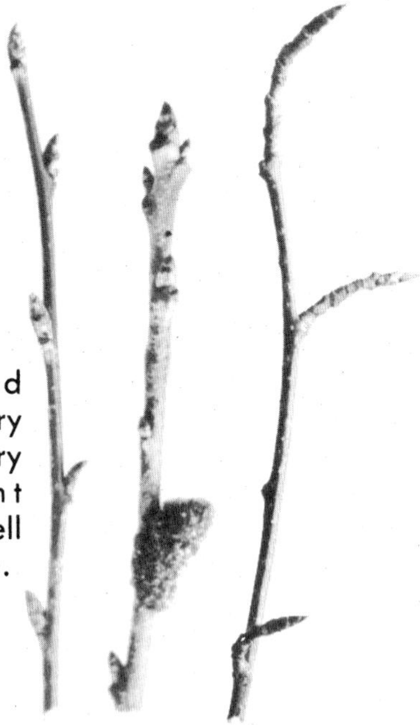

Cherry 76

Notice tent caterpillar egg mass on middle twig. This is typical of Cherry Trees.

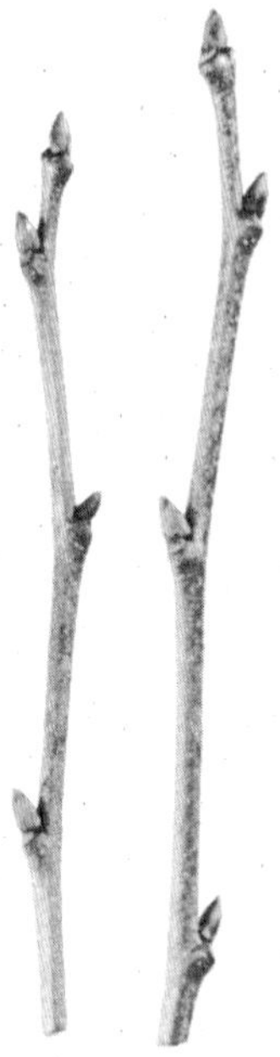

Chestnut 49

Linden 95

American Elm 42

Persimmon 115

Poplar 28

Hop Hornbeam 51

Hornbeam 50

Willow 25

Poplar 28

Magnolia 64

Mountain Ash 74

typically
zigzag

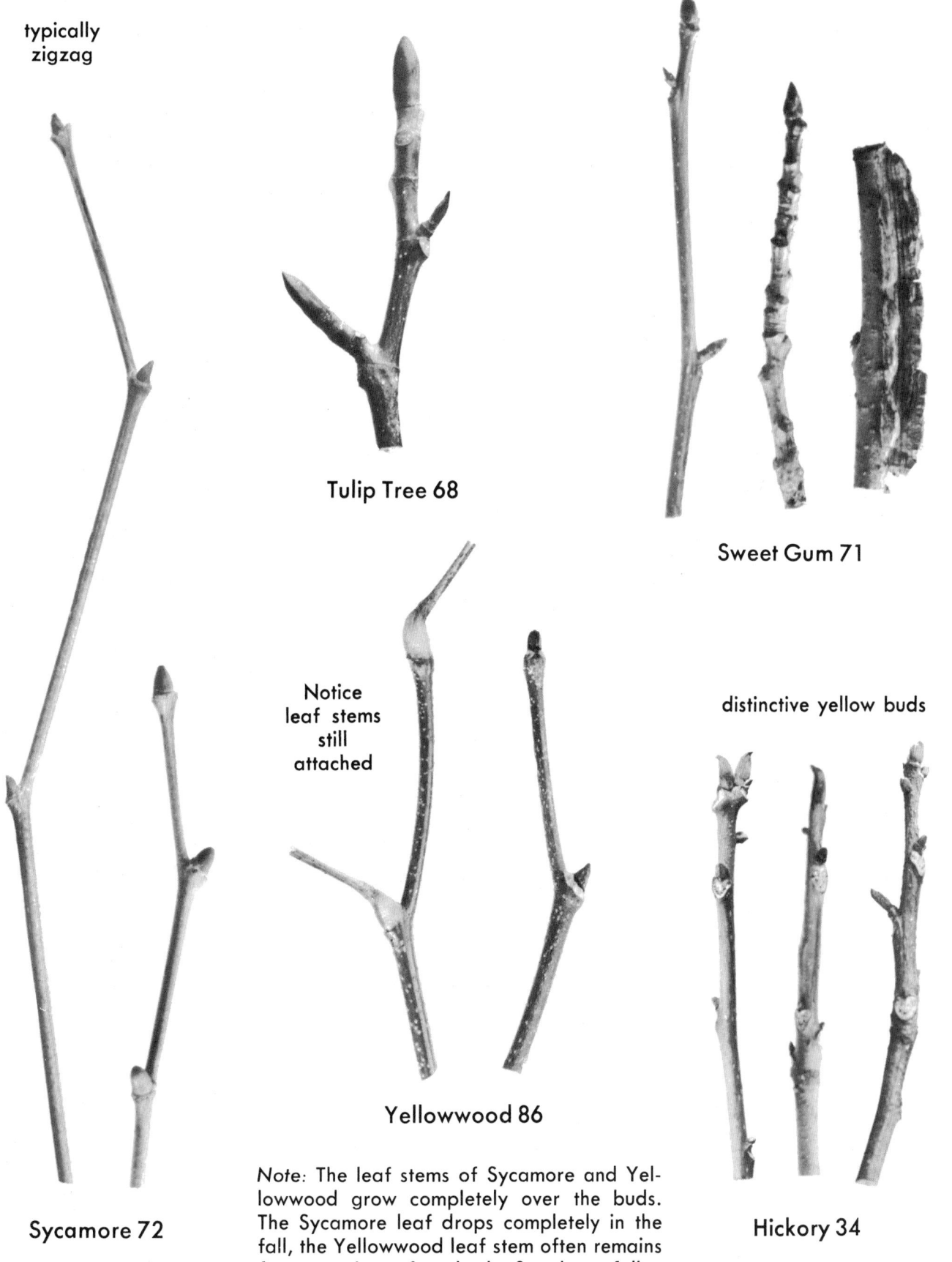

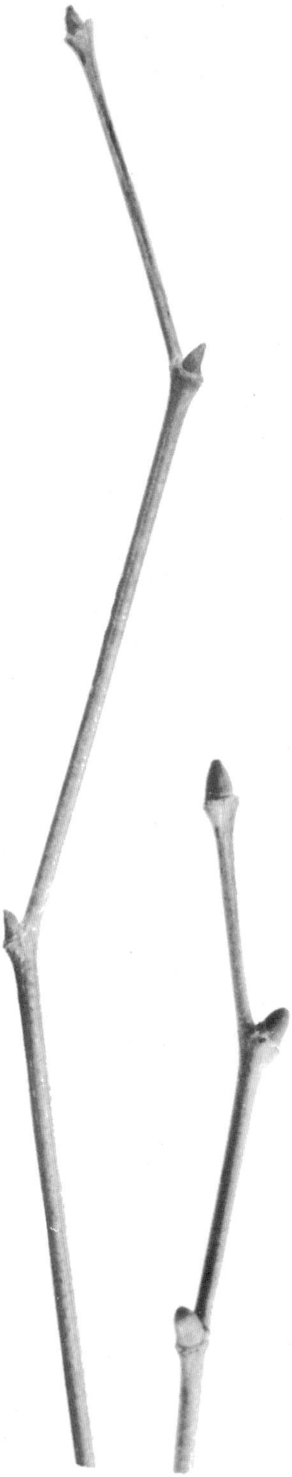

Sycamore 72

Tulip Tree 68

Sweet Gum 71

Notice
leaf stems
still
attached

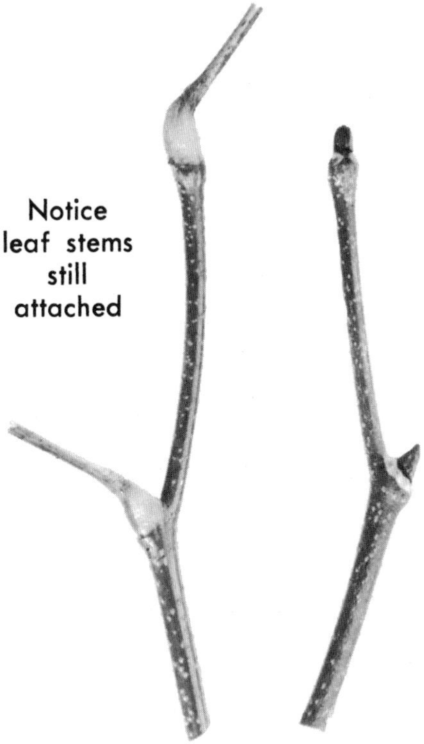

Yellowwood 86

Note: The leaf stems of Sycamore and Yellowwood grow completely over the buds. The Sycamore leaf drops completely in the fall, the Yellowwood leaf stem often remains for some time after the leaflets have fallen off it.

distinctive yellow buds

Hickory 34

Black Locust
81

Honey Locust
82

Hop Tree
88

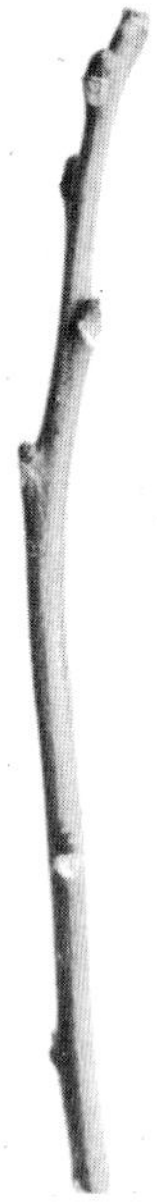

Sourwood
114

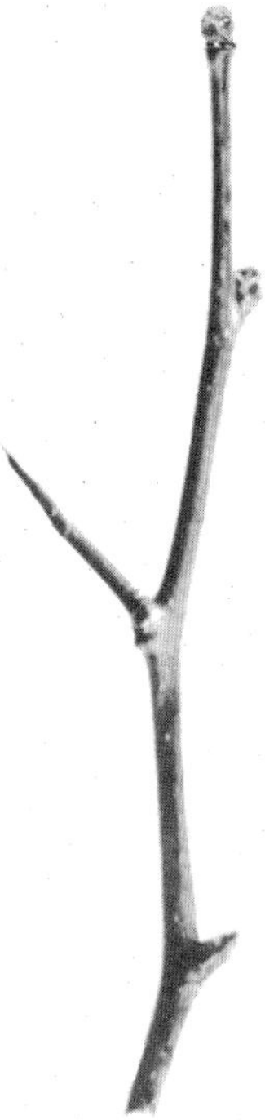

Hawthorn
73

Hackberry 41

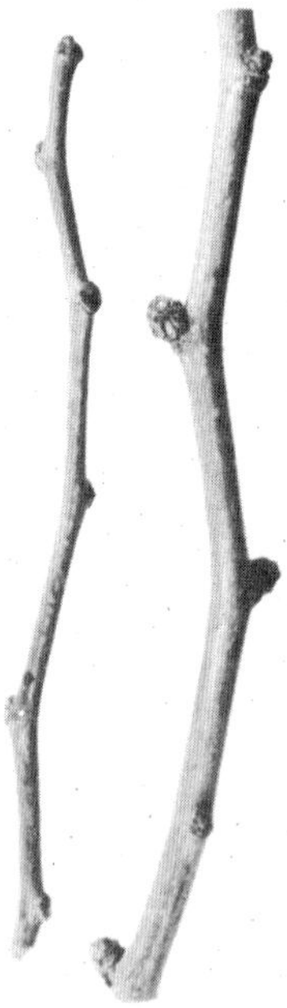

Osage Orange 62

Redbud 80

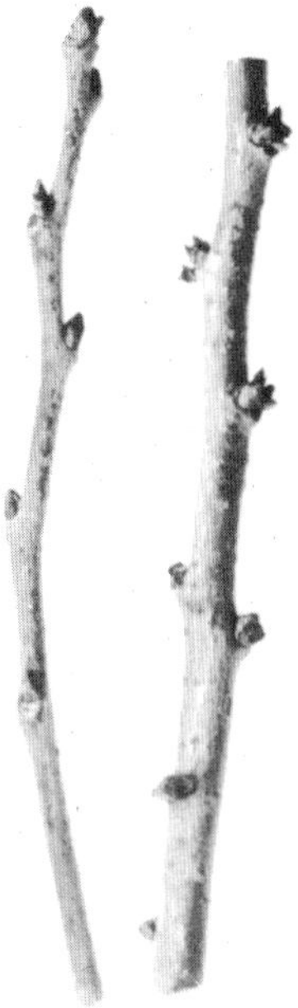

Persimmon 115

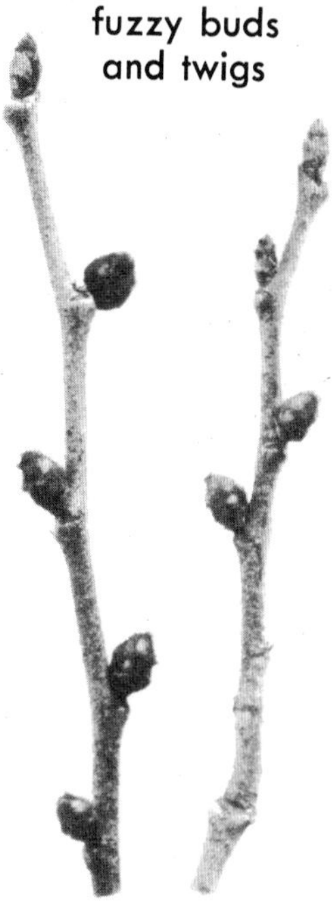

Slippery Elm 42

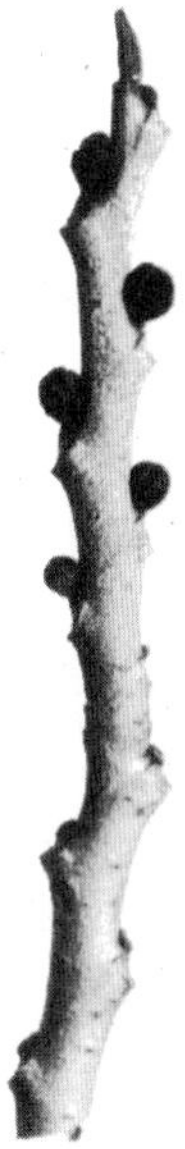

Pawpaw 63

Willow 25

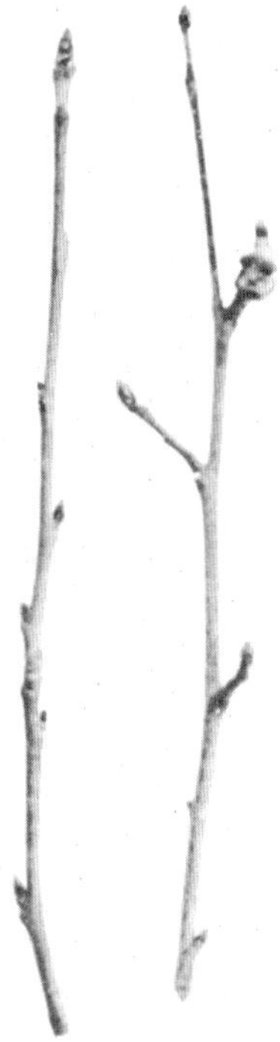

Cherry 76

Mulberry 60

Winter Catkins

These winter catkins are a form of flower bud, which elongate in spring and become flowering catkins. See Key #4 Flowers. Note, however, that although other trees bear flowers in the form of catkins, only these trees show this type of bud in winter.

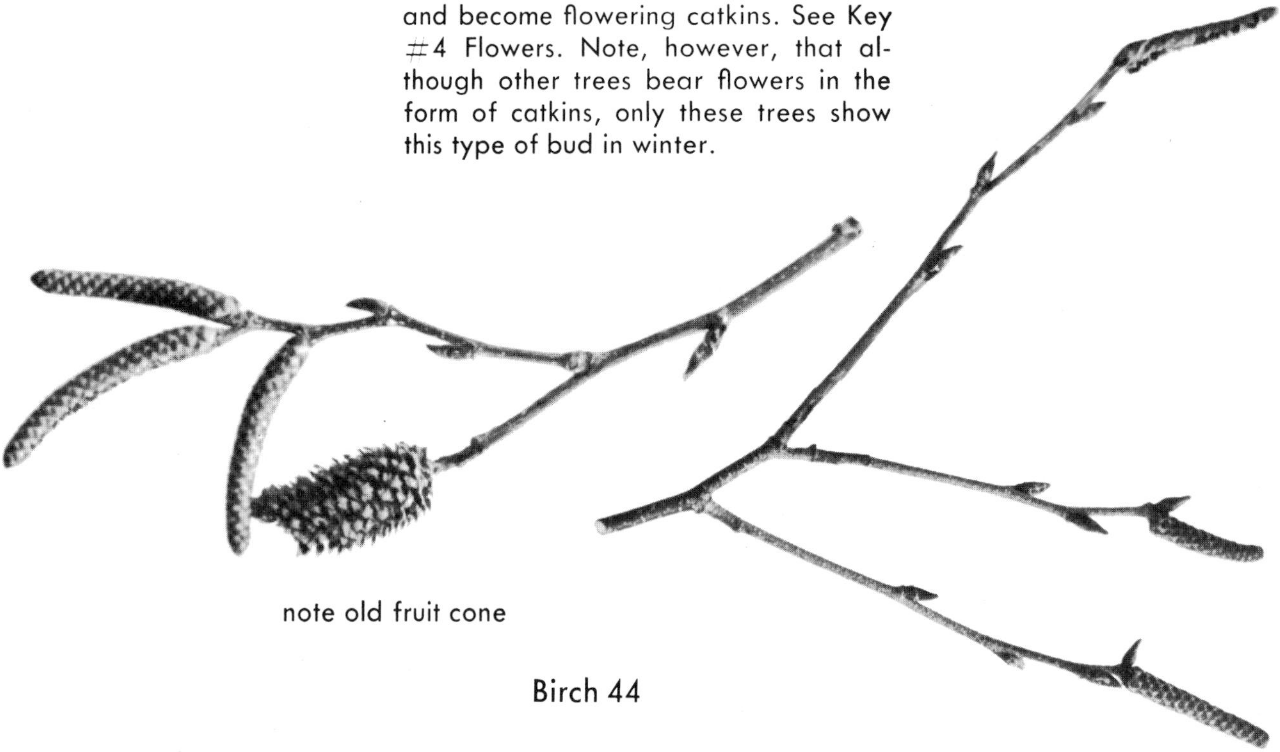

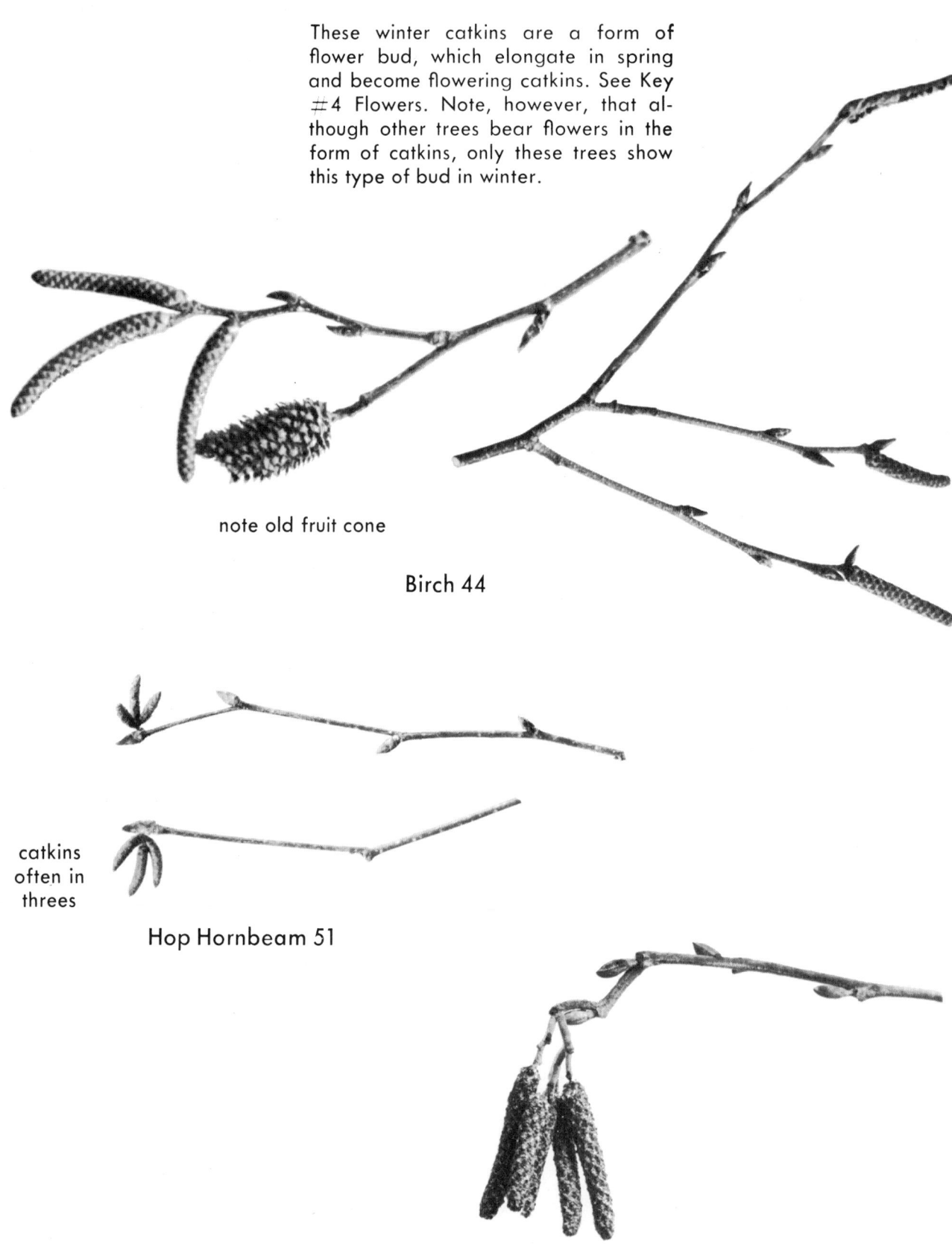

note old fruit cone

Birch 44

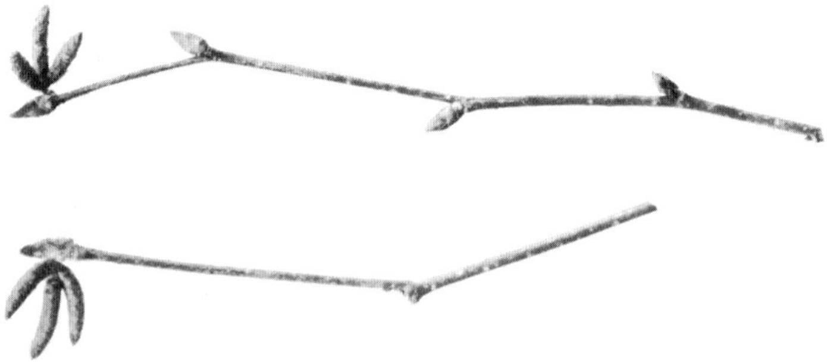

catkins often in threes

Hop Hornbeam 51

Alder (not included in rest of book as it is a shrub, but shown here to avoid confusing catkins)

Holly 94

(only broad-leaved Evergreen tree in North)

these are flower buds

typical of twig that did not produce flower bud

Paulownia 122

Ginkgo 24

Some Buds Which Swell in Late Winter

Pussy Willow 25

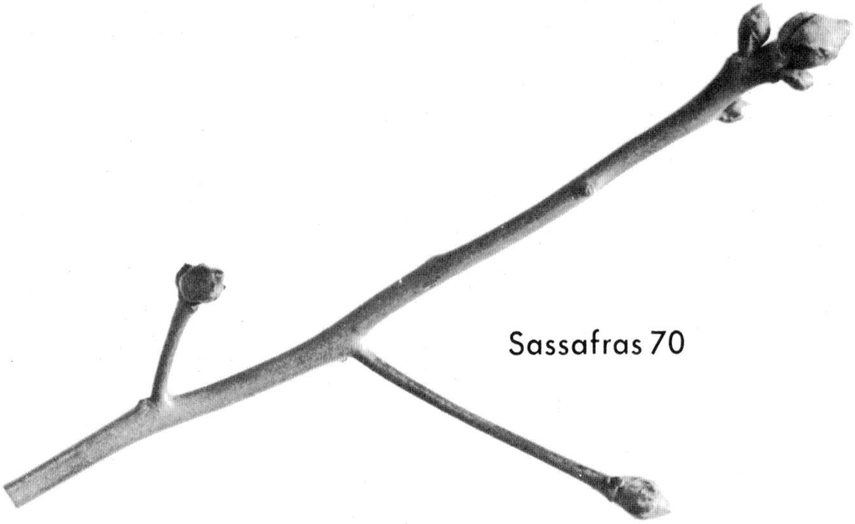

Sassafras 70

Willow 25

Maple 96

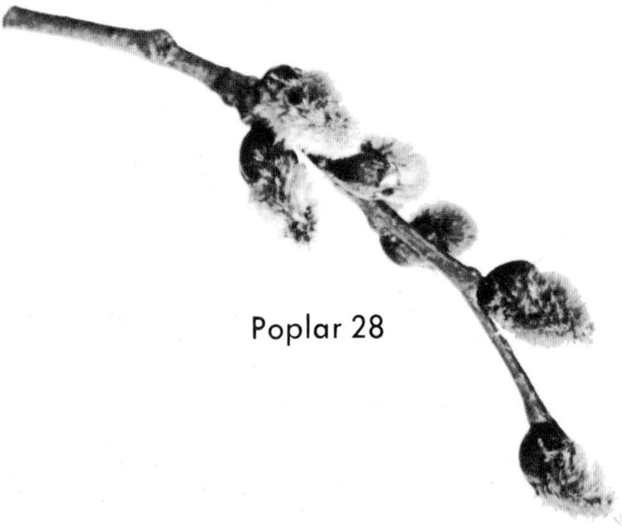

Poplar 28

Key #7 **BARK**—All pictures in same relative scale

This Key shows two pictures of each bark, in most cases, one a close-up, the other a larger section of the trunk. This has been done to approximate the composite image created by the eye. Always look at the whole trunk to get an over-all impression of the typical bark, as well as for details seen at eye level.

Notice that the mature bark of the lower trunk is often different from the young bark of upper branches. The barks selected for these pictures represent average mature barks, unless otherwise noted. The size of scales or furrows tend to become exaggerated with age, so make allowance for this in looking at very old trees.

The same kind of tree frequently has several types of mature bark, and to show this, pictures of different barks for the same trees are placed on different pages according to type. It is important, therefore, always to look for the correct type and not just for a name.

Shagbark Hickory 34

Sugar Maple 96

Red Birch 44

Black Willow 25

Black Birch 44

Yellow Birch 44

Tupelo 110

Persimmon 115

Hackberry 41

Ohio Buckeye 106

Red Maple 96

White Oak 52

Chestnut Oak 52

White Poplar 28

Tupelo 110

Cottonwood 28

Black Locust 81

Tulip Tree 68

Crack Willow 25

White Willow 25

Mulberry 60

Osage Orange 62

BK
4

Ginkgo 24

Mockernut Hickory 34

Sassafras 70

American Elm 42

Black Walnut 38

Butternut 38

Tupelo 110

Swamp White Oak 52

Linden 95

Red Ash 116

Large-toothed Poplar 28

Tulip Tree 68

Norway Maple 96

White Ash 116

Catalpa 124

Sweet Gum 71

Slippery Elm 42

Burr Oak 52

Bitternut Hickory 34

Balm of Gilead 28

Sourwood 114

Pignut Hickory 34

Tulip Tree (young) 68

White Oak 52

BK 8

Hickory (young) 34

Balsam Poplar 28

Willow Oak 52

Pin Oak 52

Scarlet Oak 52

Ailanthus 89

Honey Locust 82

Sweet Buckeye 106

Tupelo 110

Black Ash 116

Sugar Maple 96

Paulownia 122

Sugar Maple 96

Red Birch 44

Red Oak 52

Black Walnut 38

Sweet Gum 71

Magnolia 64 (Cucumber Tree)

Kentucky Coffee Tree 84

White Oak 52

Catalpa 124

Silver Maple 96

Horsechestnut 106

Red Maple 96

Black Birch 44

Blue Ash 116

Black Cherry 76

Black Oak 52

Sycamore Maple 96

American Elm 42

Hop Hornbeam 51

White Oak 52

Tupelo 110

Nannyberry 127

Flowering Dogwood 112

Hop Hornbeam 51

Hop Tree 88

Redbud 80

Sycamore 72

Red Maple 96

Pecan 34

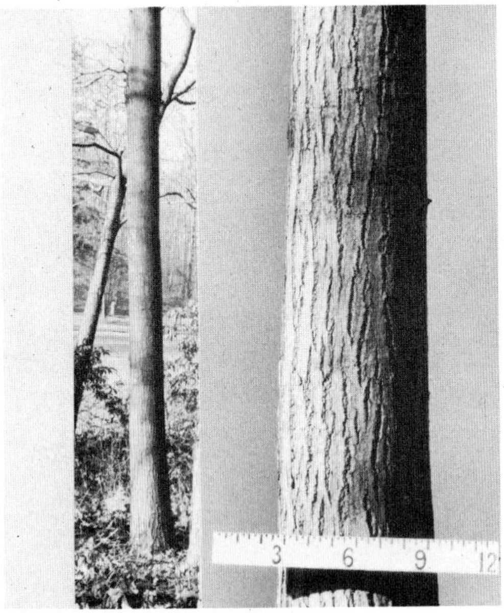

Chestnut (young) 49

Sugar Maple (young) 96

Ailanthus 89

Hackberry 41

Pawpaw 63

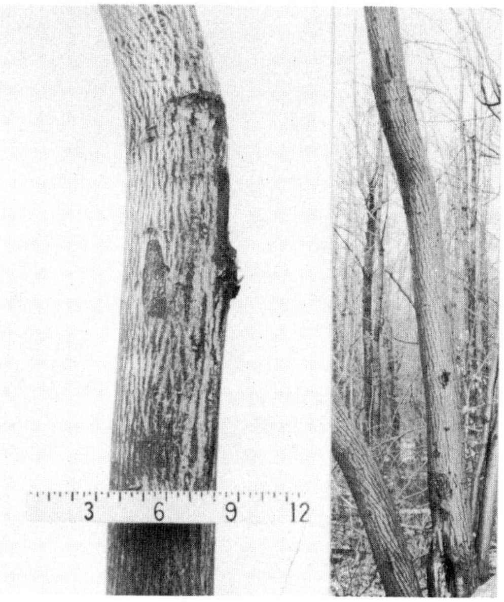

Shadbush 79

Hornbeam 50

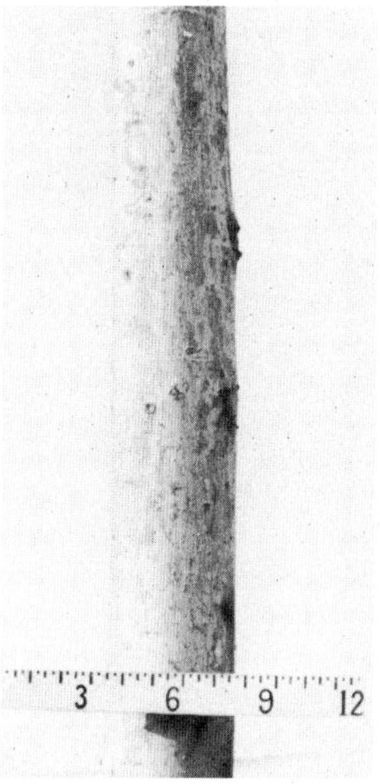

Red Maple (young) 96

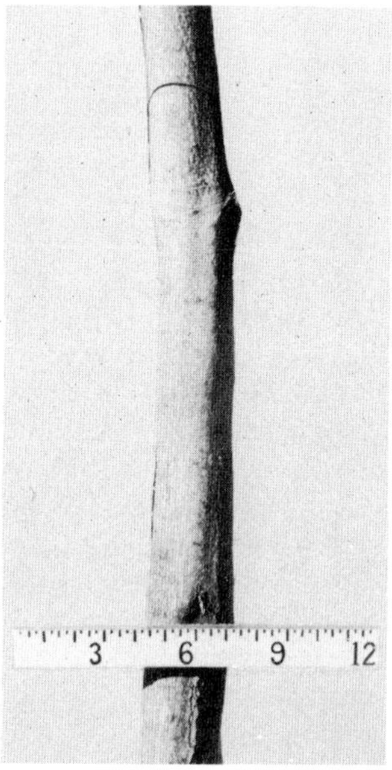

Alt-leaved Dogwood 112

Beech 48

Yellowwood 86

Paper Birch 44

Gray Birch 44

American Mountain Ash 74

Sweet Cherry 76

European Mountain Ash 74

Magnolia 64 (Umbrella Tree)

Large-toothed Poplar 28

Trembling Aspen 28

Trembling Aspen 28

Pin Cherry 76

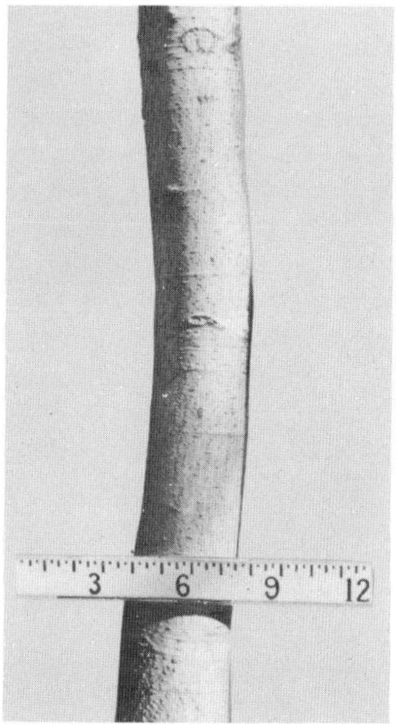

Magnolia 64 (Sweet Bay)

Hawthorn 73

Yellow Birch 44

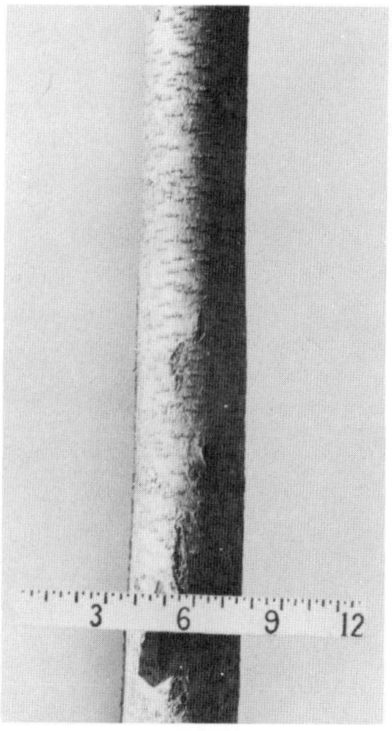

Sumac 90

Black Birch (young) 44

NEEDLE-LEAVED TREES

Master Key of Genera. The Species are identified in the Master Pages.

ALL ARE EVERGREEN except Bald Cypress and Larch which are deciduous (losing leaves in fall).

ALL ARE CONE-BEARING except Yews which have red berries. The Junipers appear to have berries, but their fruit is actually a cone, the scales of which grow together.

Master Key of Genera—All actual size

Refer to the pictures while reading this.

NEEDLES GROWING IN CLUSTERS—each cluster growing from one place on the twig

A. In clusters of 2, 3 or 5—Pine (evergreen)

B. Many in a cluster—Larch (deciduous)

NEEDLES GROWING SINGLY—only one needle growing from the same spot on the twig

A. Four-sided needles—born on short projections on twig, which can be felt and seen where needles have fallen off older twigs—Spruce

B. Flat needles

1) Fir:
Upright cones (see Master Pages for cones)
Rounded buds
Leaf scar round, distinct, especially on the older wood
Needles usually curved in one direction on each side of twig
Needles usually whitish on underside (except a few such as White Fir, a western tree, which has very long needles for a fir, pale bluish or whitish green on both sides)

2) Douglar Fir:
Hanging cones
Sharp-pointed buds
Leaf scar oval, not so distinct as Firs
Needles fairly straight and growing all around twig uniformly
Needles nearly uniform in color on both sides, sometimes lighter underneath

3) Yew:
Red berries
Needles dark green above, yellow-green underneath

4) Hemlock:
Cones
Needles whitish underneath

5) Bald Cypress:
Round (ball-like) cones
Needles very soft, short and feathery
Deciduous

SCALELIKE OR PRICKLY NEEDLES

A. Either type—Sometimes both on same tree—Red Cedar
Light blue, berrylike fruit
Needles when scalelike feel four-sided and look so under a magnifying glass

B. All scalelike

1) White Cedar:
Cones small, round, somewhat indistinct
Needles flattish (not four-sided like Red Cedar)

2) Arborvitae:
Cones distinct
Needles very flat

NEEDLES GROWING IN CLUSTERS

for Pine cones see Master Pages

Notice needles grow out of distinct sheath (from one place on twig).

Pine (evergreen) 1

A. Needles in clusters of: 2 3 5

Larch cones vary in size depending on species, but all kinds remain on trees for a long time. (See Master Pages for pictures.)

Larch (deciduous) 10

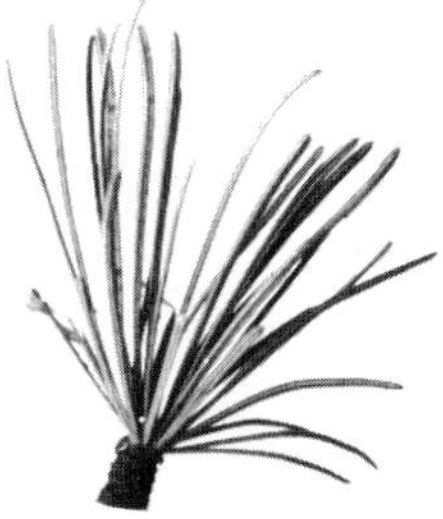

B. Needles many in a cluster

winter twigs

Needles Growing Singly

A. Four-sided needles — Spruce

Borne on short projections on twig. These projections can be felt and seen where needles have fallen off.

B. Flat Needles

Fir
Douglas Fir
Yew
Hemlock
Bald Cypress

Spruce 12

Four-sided needle can be felt by turning a single needle between thumb and forefinger, or is clearly seen with a magnifying glass. Test the end nearest the twig; the pointed end may look three-sided. Spruce needles are stiff and sharp (Fir and Douglas Fir feel soft when the hand is pulled outward along individual twig; Spruce, stiff and sharp).

notice rounded buds

Fir 16

Cones are upright,
buds are rounded,
leaf scar *round,* distinct (look inside on older twigs where needles have fallen off)—very different from projections on Spruce twigs.
Needles are curved.
Needles of two eastern Firs are whitish underneath, as are most Firs. There are a few exceptions, such as the western White Fir which has uniform, light-colored needles, much longer than most Firs.

notice sharp-pointed buds

Douglas Fir 15

Cones hang down,
buds are sharp-pointed,
leaf scar *oval,* but less distinct than on Firs.

Needles are fairly straight.
Needles fairly uniform in color on both sides. (Sometimes lighter underneath.)

red berries

yellow-green beneath

Yew 23

cone

Hemlock 18

whitish beneath
There are two white lines (stomatic bands) on underside of needle.

ball-like cones

soft, short feathery needles

Bald Cypress 19
deciduous

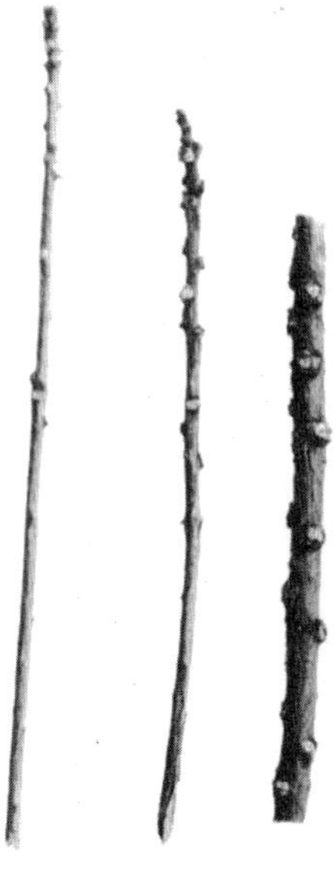

winter twigs

Scalelike or Prickly Needles

A. Either type—sometimes both on same tree

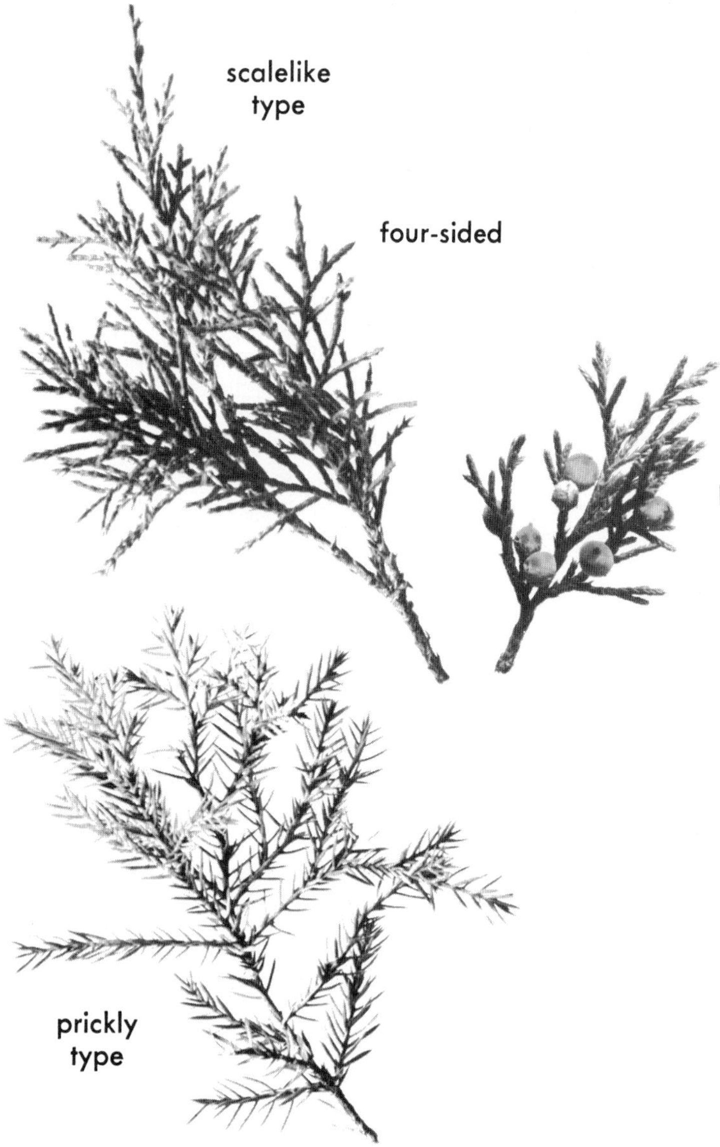

Red Cedar 20
(Juniperus virginiana)

These two types of needles often appear on the same tree, and when found are sure sign of Juniper. If only scaly type is found, it can be told by four-sided needles which can be seen, or better, felt.

Note: Common Juniper *(J. communis)*, usually a shrub, has *only* prickly needles.

B. All Scalelike

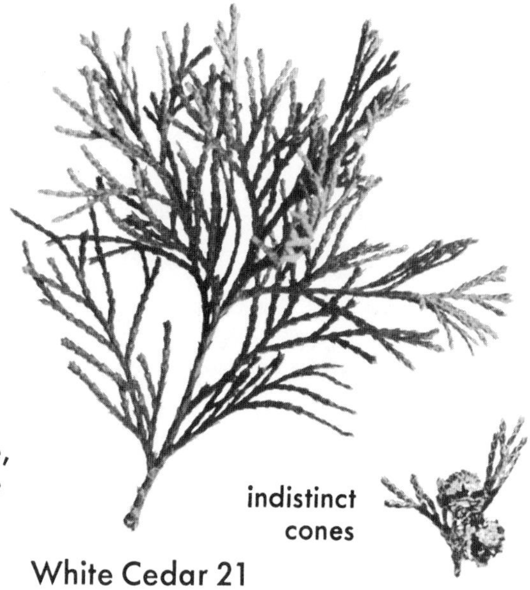

White Cedar 21
(Chamaecyparis thyoides)

Note: White Cedar has flat appearance and feel, unlike four-sided Red Cedar. Arborvitae is very flat, as pictured below.

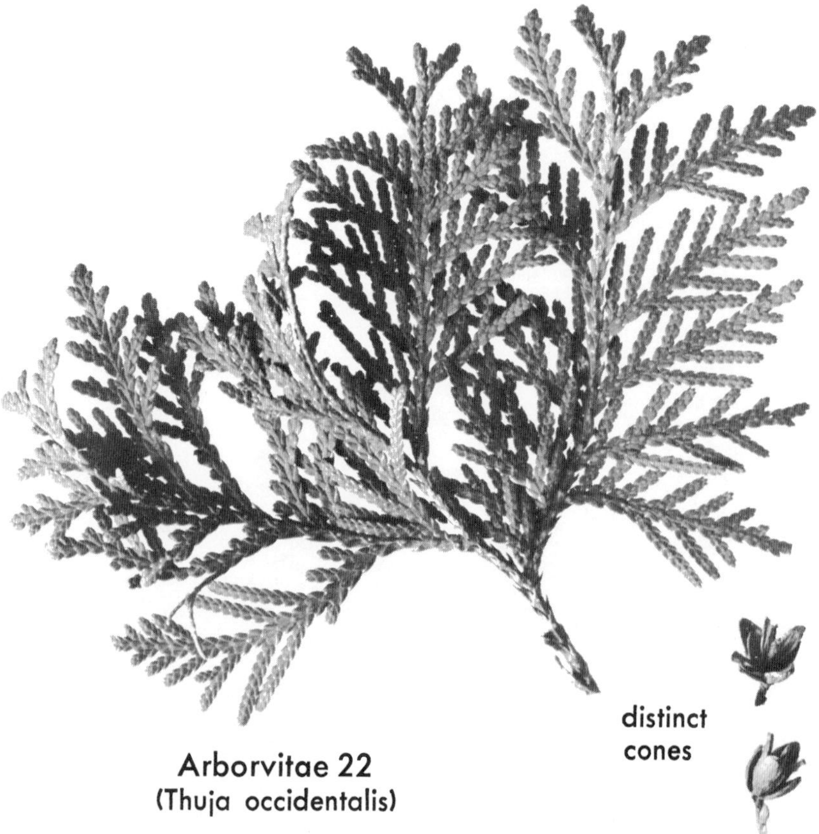

Arborvitae 22
(Thuja occidentalis)

PART II

MASTER PAGES

NEEDLE-LEAVED TREES
ALL ARE EVERGREEN except Bald Cypress and Larch.

BROAD-LEAVED TREES
ALL ARE DECIDUOUS except Holly which is evergreen.

MASTER PAGES

The Master Pages have two main uses: (1) To assemble in one place the important features of each tree, and (2) To identify species. The genera are identified in the Keys. If the genus is known, it is not necessary to consult the Keys, and the Master Pages can be turned to directly for species identification by consulting the index in the back of the book.

The Master Pages sometimes show only one flower, or other detail, for a particular genus, and in such cases it can be assumed that the detail is typical of the genus and therefore does not assist in determining the species.

At the heading of each Master Page the common name is given, followed by the botanical one. This is followed by a flower note: whether perfect (both male and female parts in the same flower), monoecious (distinct male and female flowers on the same tree), or dioecious (male flowers on one tree, female flowers on a different tree). The date of blooming is given, but obviously will not be exact for every part of the range, but can be used to determine the sequence of bloom for any area. Comparison with a few known dates of bloom will indicate whether the dates in the book are correct for a given area, or if not, how much to add or subtract.

Following the information on flowers, the size of the trees is noted: small, medium, medium to large, and large. This information is more useful for purposes of recognition than for identification, because it is part of knowing a tree at various ages, while identification is based on details which remain fairly constant regardless of the size of a tree.

Next, information is given concerning where the trees are most likely to be found. However, as explained in the introduction to this book, trees may be encountered far from their original range, so never eliminate one for purely geographical reasons.

Each Master Page features first a picture of the whole tree or trees. This is followed by specific details, and in cases where several species are included in one genus, comparison of these details leads to species identification. As explained before, it is by comparing all the details that identification is achieved, and the combination of determining factors is not the same in each case. For example, it may be bark and twig for one; leaf, fruit or flower for another. Throughout the Master Pages, notes point out important differences between trees, which otherwise might be difficult to identify.

PINES – *Pinus*

Flowers Monoecious—May-June Cones usually mature in two years (sometimes three)

Austrian Pine	*Pinus nigra*	large	From Europe—generally planted here
Jack Pine (Gray Pine)	*P. banksiana*	medium	Almost only northern states
Loblolly Pine (Southern Yellow or Old-field Pine)	*P. taeda*	large	Mostly southern states not far from coast, north to New Jersey
Long-leaf Pine (Southern Yellow Pine)	*P. palustris*	large	Southern states, mostly in coastal plains
Pitch Pine	*P. rigida*	medium to large	Northeastern states and Appalachians
Red Pine (Norway Pine)	*P. resinosa*	medium to large	Mostly northern states
Scots Pine	*P. sylvestris*	large	From Europe—generally planted here
Short-leaf Pine (Southern Yellow Pine)	*P. echinata*	large	Southern states; found north to New York
Table Mountain Pine	*P. pungens*	small to medium	Found mostly in Appalachians
Virginia Pine (Jersey or Scrub Pine)	*P. virginiana*	medium	Mostly across middle of range
White Pine (Eastern White Pine)	*P. strobus*	large	Mostly northern states and Appalachians

A special Key is used to identify the above Pines. This is based on two characteristics of their needles:

(1) The number of needles growing from a single cluster (or sheath)

(2) The length of the needles

The table below gives this in outline form, and pages 4 and 5 show pictures of the actual needles.

needles in a cluster	¾ — 1½″	1½ — 3″	3 — 6″	6″ and longer
5			White Pine	
3			Pitch Pine	Loblolly Pine Long-leaf Pine
2 and 3		Table Mountain Pine (usually only 2)	Short-leaf Pine (usually both 2 & 3 on same tree)	
2	Jack Pine	Scots Pine Virginia Pine	Austrian Pine Red Pine	

To verify identification made by the above method, or in cases of doubt, look for cones (usually on the tree or beneath it).

Jack Pine

Virginia Pine

Scots Pine

Red Pine

Austrian Pine

Short-leaf Pine

Table Mountain Pine

White Pine

Pitch Pine

Loblolly Pine

Long-leaf Pine

2-needle clusters

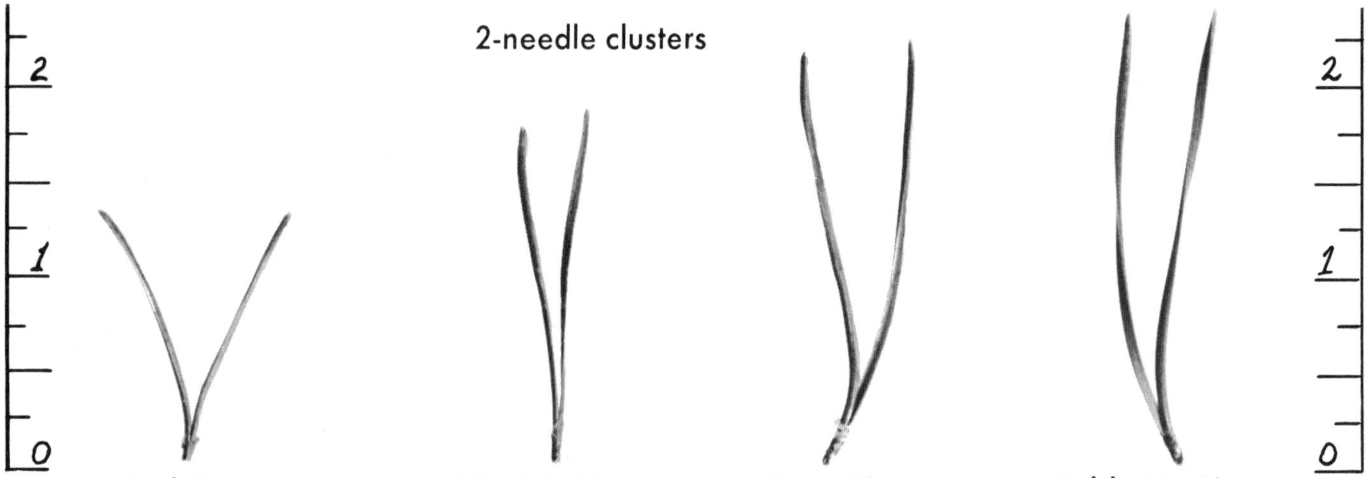

Jack Pine
¾ — 1 ½″

Virginia Pine
1 ½ — 3″

Scots Pine
1 ½ — 3″
(bluish cast)

Table Mt. Pine
1 ½ — 3″
(usually 2 needles occasionally 3 needles)

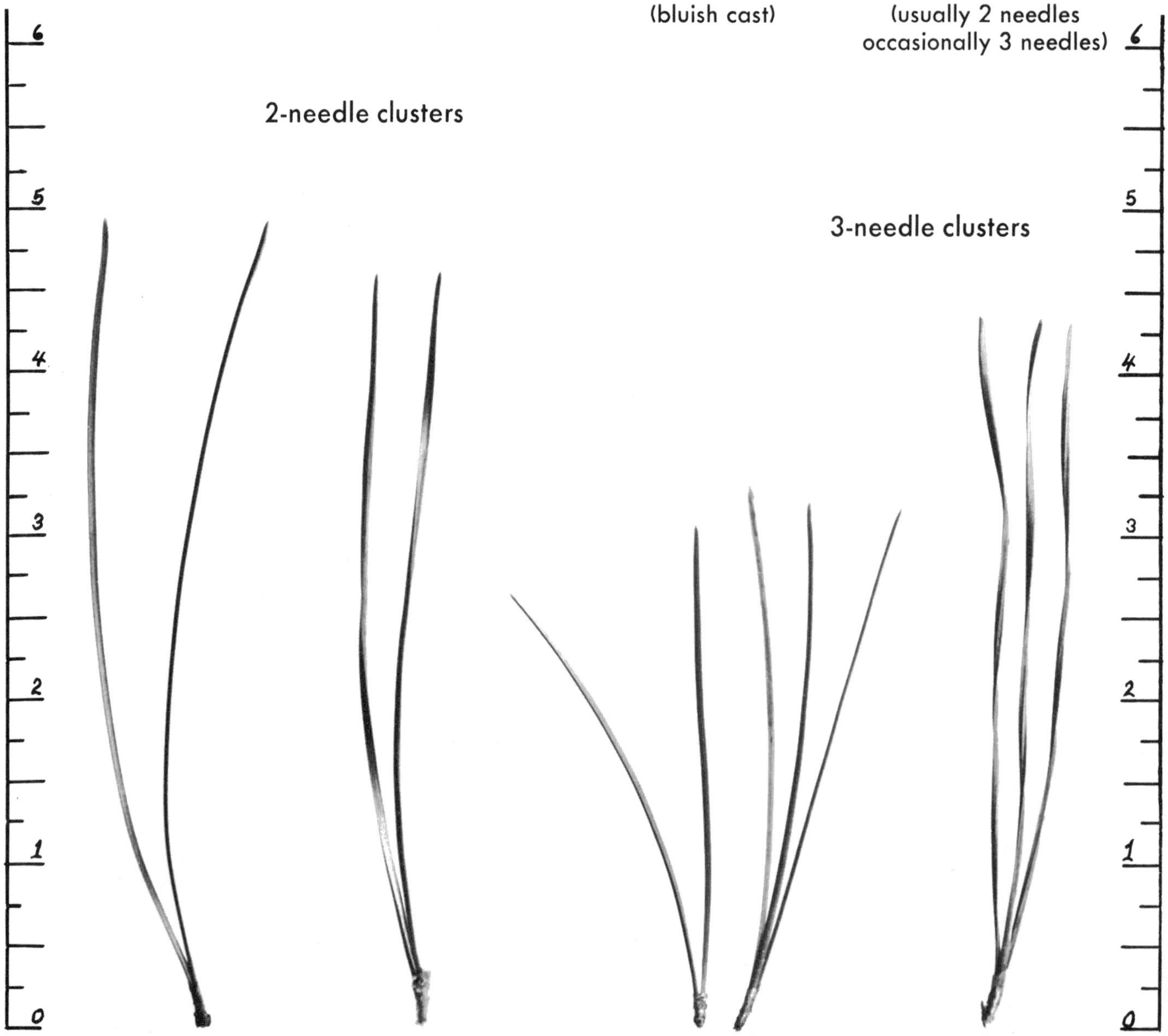

Red Pine
4 ½ — 6″ (occasionally longer)
(flexible needles)

Austrian Pine
3 ½ — 6″
(sharp, stiff needles)

Short-leaf Pine
3 — 4 ¾″
(usually both 2- and 3-needle clusters on same tree)

Pitch Pine
3 — 5 ½″
(only 3-needled Pine found in the north)

Note: The average variation of needle length is given under each picture. Be sure to look for an average needle, as every tree may have some needles longer or shorter than indicated. Occasionally an individual tree may vary widely from the average, but this is exceptional.

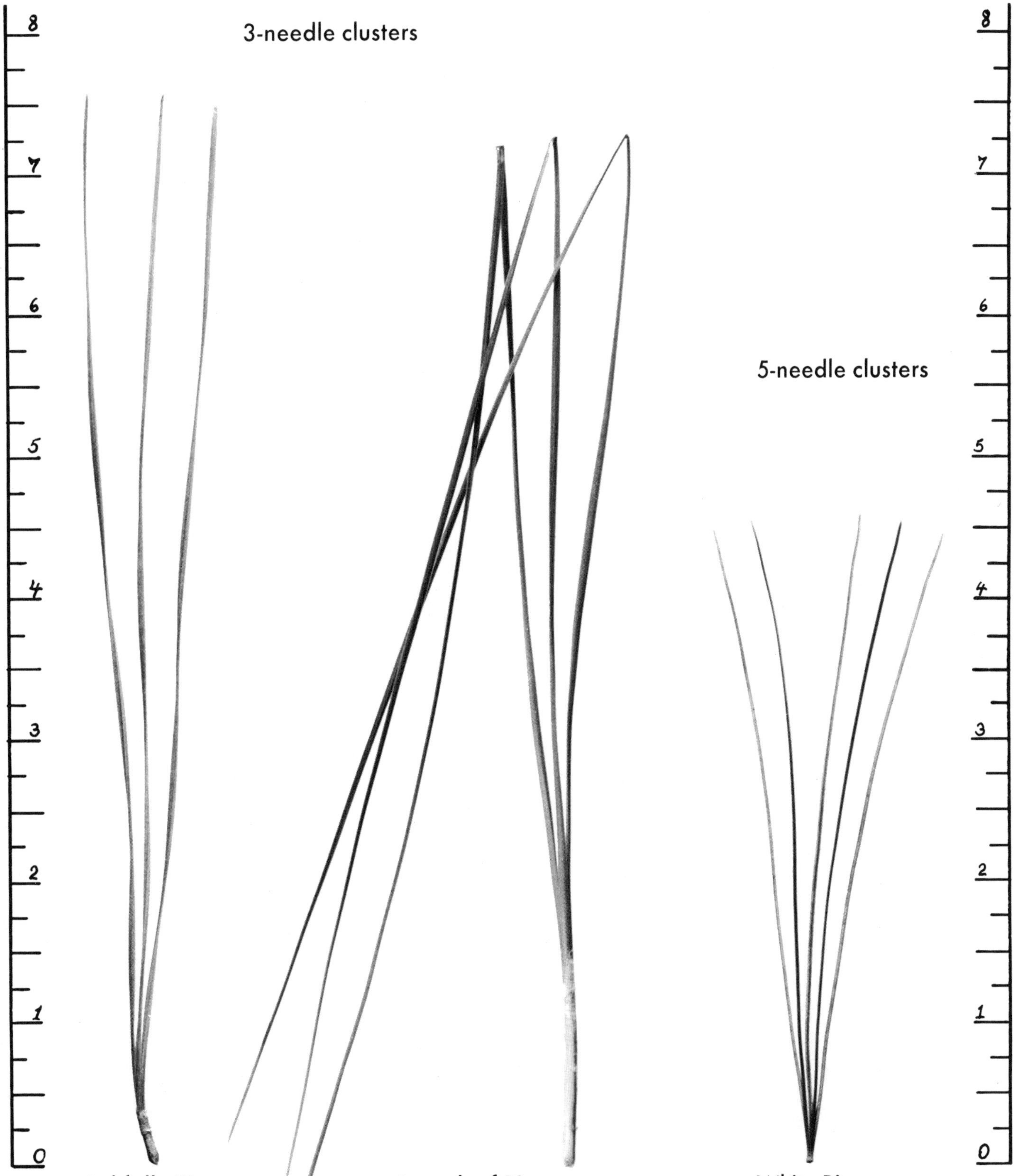

Loblolly Pine
6 — 9″

Long-leaf Pine
8 — 18″

White Pine
2¼ — 5½″

SCALE
ACTUAL SIZE

Jack Pine
1 — 2″

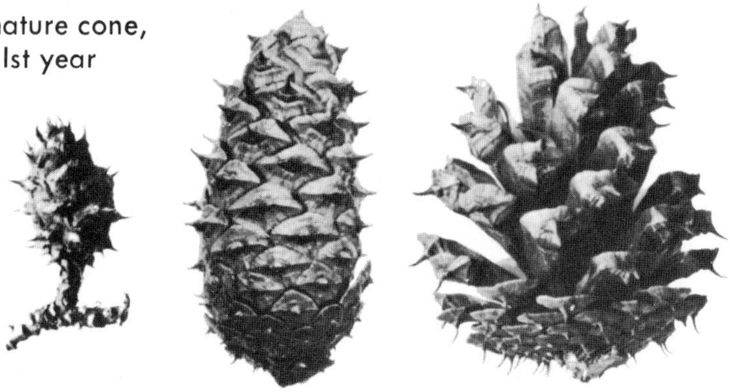

Virginia Pine
1½ — 2½″

Scots Pine
1 — 2½″

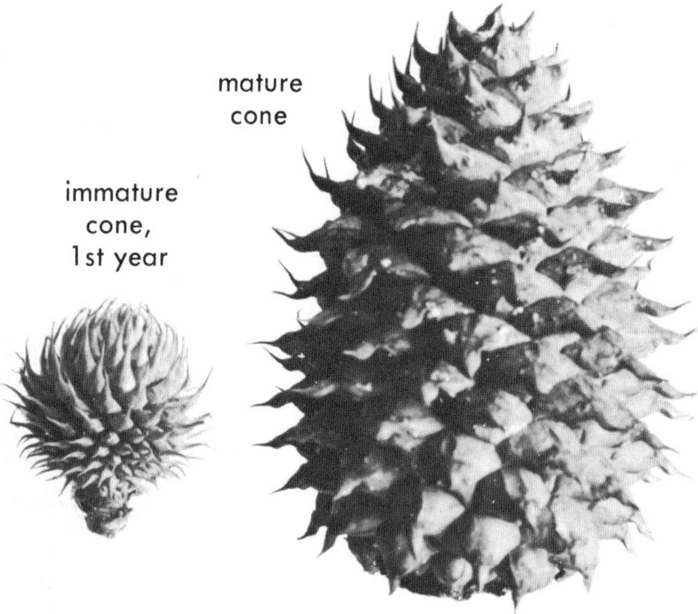

Table Mountain Pine
2 — 3½″

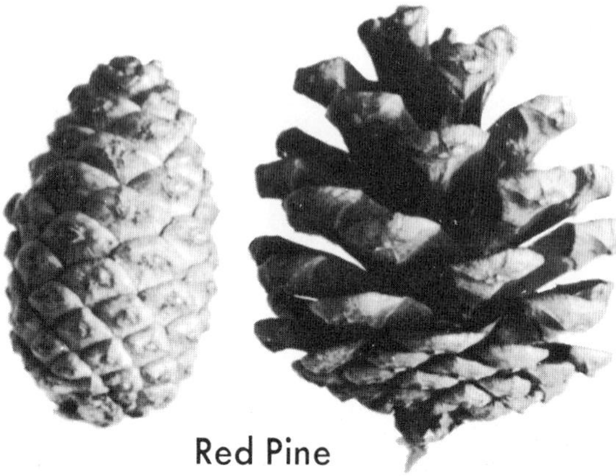

Red Pine
1½ — 2¼″

Austrian Pine
2 — 3½″

All Pine cones take 2 years to mature (occasionally 3 years). They then open, dispersing winged seeds.

Average variation of mature-cone length indicated under each name.

SCALE
ACTUAL SIZE

mature cones
unopened open

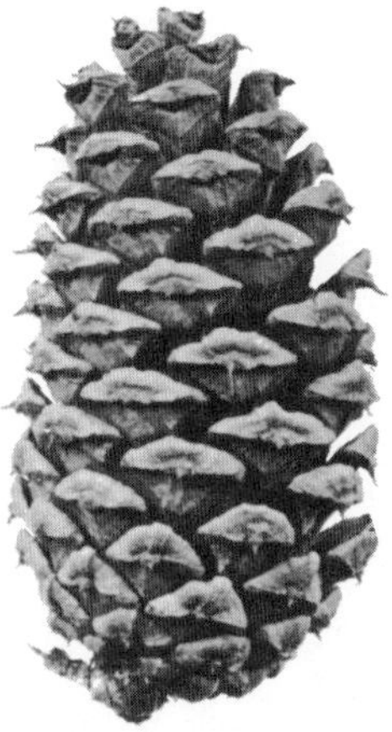

immature cone, 1st year

Short-leaf Pine
1 ½ — 2 ½″

mature cone

immature cone, 1st year

White Pine
4 — 8″

mature cones
unopened open

immature cone, 1st year

Pitch Pine
2 — 3 ½″

See next page for cones of Loblolly and Long-leaf Pine

SCALE
ACTUAL SIZE

Long-leaf Pine
6 — 10″

Loblolly Pine
3 — 6″

mature unopened

mature open

SCALE
ACTUAL SIZE

female (pistillate)
flowers at end
of new growth

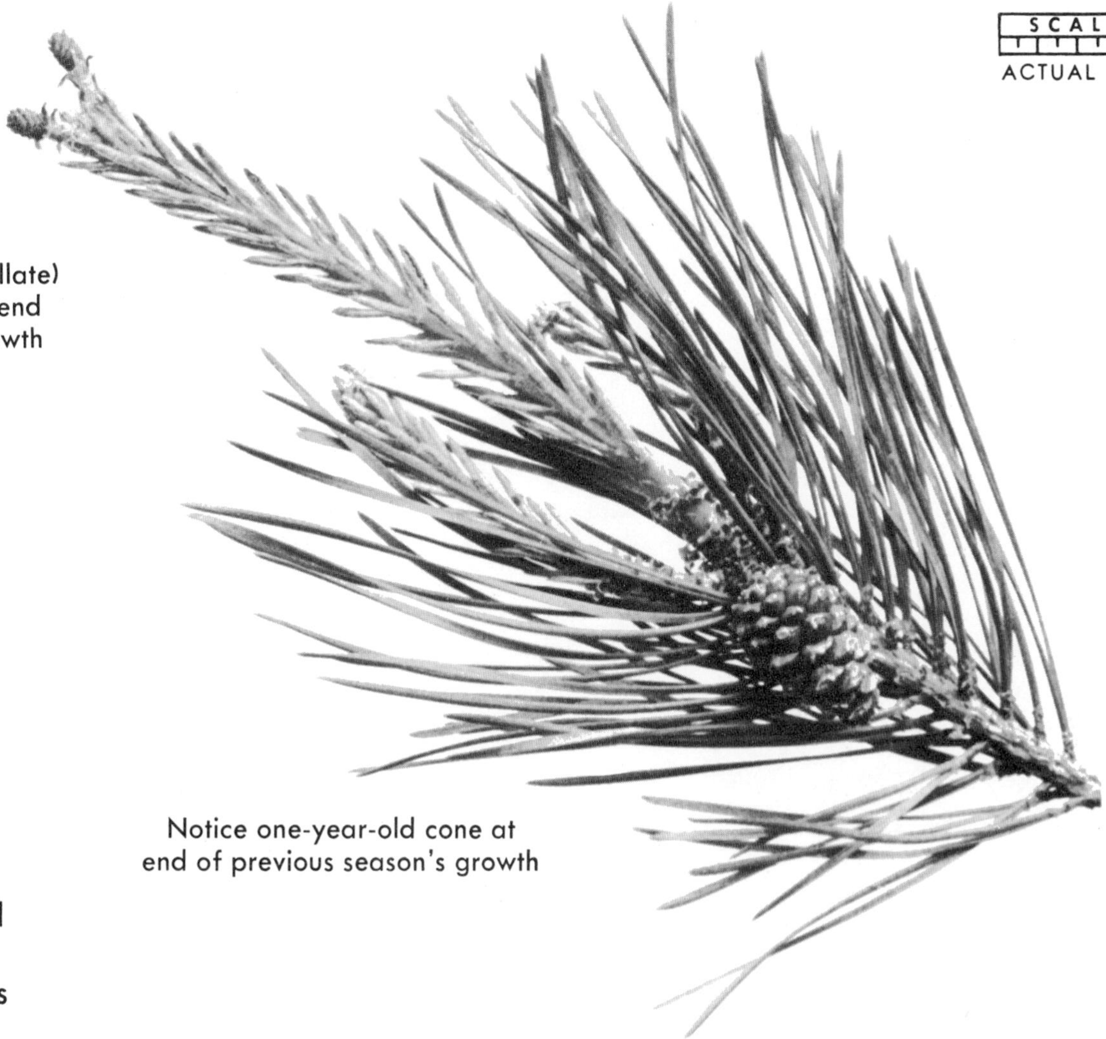

Notice one-year-old cone at
end of previous season's growth

Typical Pine Flowers

male (staminate)
flowers along new growth,
covered with pollen

LARCH—*Larix*

Flowers Monoecious — May

Deciduous Trees
(losing needles in fall)

American Larch (Tamarack or Hacmatac)	*Larix laricina*	medium	Found largely in northern states
European Larch	*L. decidua*	large	From Europe—will grow farther south than American Larch

European Larch

American Larch

Larch in summer

SCALE
ACTUAL SIZE

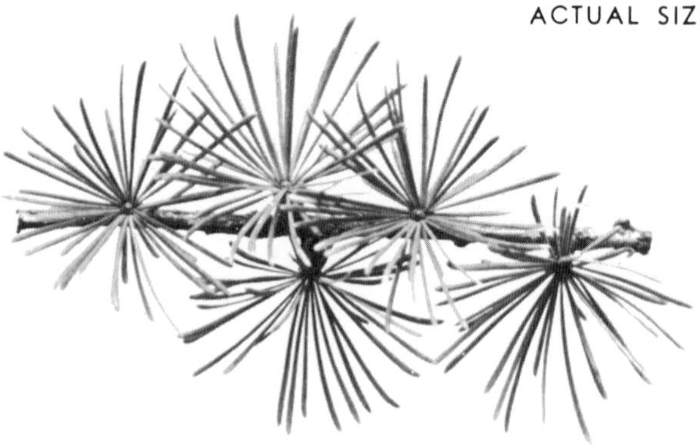

Larches produce many needles in a cluster. The Bald Cypress is the only other deciduous needle-leaved tree in our range (see M.P. 19).

Typical Larch
Flowers

SCALE
ACTUAL SIZE

European Larch

winter twigs

American Larch

European Larch
(large cones)

American Larch
(small cones)

SPRUCE—*Picea* Flowers Monoecious—May

Black Spruce	*Picea mariana*	large	Northern states
Colorado Spruce	*P. pungens*	medium to large	Comes from western U. S.
(The well-known Blue Spruce is a variety of the Colorado Spruce.)			
Norway Spruce	*P. abies (or P. excelsa)*	large	From Europe, widely planted in U. S.
Red Spruce	*P. rubens*	medium to large	Northeast and Appalachians
White Spruce	*P. glauca*	large	Northern states; most northerly growing Spruce

Red Spruce

Black Spruce

White Spruce

Norway Spruce

Colorado Spruce

MP 12

Note: Spruce needles are four-sided, stiff and sharp.

Red, Black and White Spruces are difficult to tell apart unless seen together, but notice:

Red Spruce is darker green and usually has more forward-pointing needles.
Black Spruce is lighter, bluish green.
White Spruce is light yellowish- to bluish-green, and needles appear further apart.

SCALE
ACTUAL SIZE

Red Spruce

Black Spruce

White Spruce

Norway Spruce

Needles appear to grow flatter (less around the stem) than other Spruces. Usually dark green. Twigs tend to be yellower than others.

Colorado Spruce

The Colorado Spruce tends to have longer needles than the others; the color varies from dark green to blue-green.

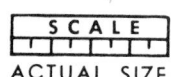

SCALE
ACTUAL SIZE

Spruce cones: average length indicated under names.

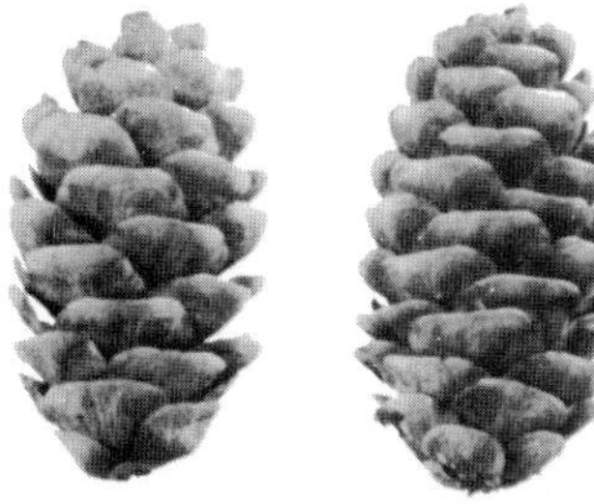

White Spruce
1 ¼ — 2″

male (staminate)

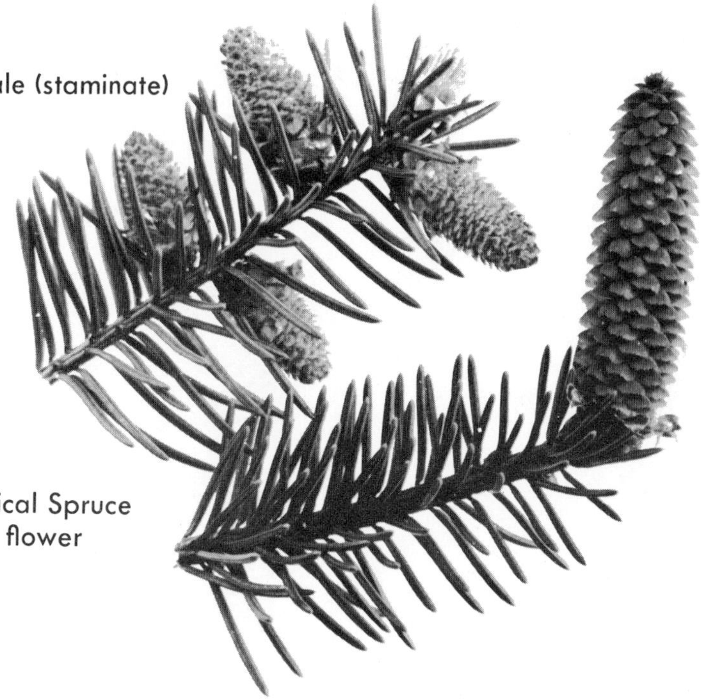

female (pistillate)

this turns into cone which hangs down

typical Spruce flower

Black Spruce
¾ — 1 ¼″

Note: White Spruce cones are more cylindrical than the oblong cones of Black and Red Spruces. Also outer ends of scales of White Spruce cones look cut off, those of Black and Red Spruces are more rounded.

Black Spruce cones tend to be the smallest, see lengths shown.

Red Spruce cones are red.

Red Spruce
1 — 1 ½″

Colorado Spruce
2 — 4″

Norway Spruce
4 — 7″

When mature, Spruce cones disperse winged seeds

DOUGLAS FIR—*Pseudotsuga taxifolia*

Flowers Monoecious—April-May

Very large (sometimes over 300′) but has not been growing in the East long enough to reach sizes found in West.

Comes from West Coast and Rockies, but now extensively planted in East.

Note: The Douglas Fir is *not* a true Fir (*Abies*). The various details are quite different. See Firs on next page.

SCALE
ACTUAL SIZE

male (staminate) flowers

female (pistillate) flowers

sharp-pointed buds (unlike the rounded ones of the true Firs)

The needles are fairly straight and uniform in color on both sides, growing more or less all around the stem.

The cones hang down, unlike the true Fir cones which stand upright. Notice the very long "bracts" coming out from between the scales, making these cones very distinctive.

FIR—*Abies* Flowers Monoecious—April-May

Balsam Fir	*Abies balsamea*	medium to large	Northern states almost exclusively
Fraser Fir (Fraser Balsam Fir or Southern Balsam Fir)	*A. fraseri*	medium	Found mostly in small area from W. Virginia to N. Georgia. Sometimes planted elsewhere.

(Neither of these Firs does well outside its natural range, although Fraser Fir is more adaptable.)

(The White Fir—*Abies concolor* from the West is often planted in the East. It has light blue-green needles.)

Balsam Fir

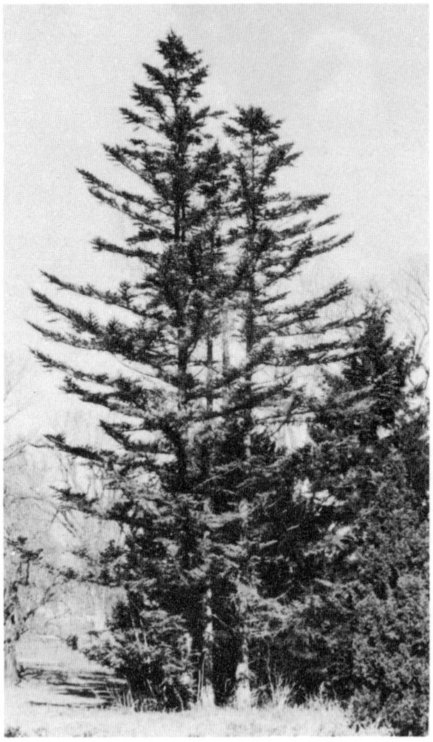

Fraser Fir

SCALE
ACTUAL SIZE

Balsam Fir

Fraser Fir

SCALE
4X ACTUAL SIZE

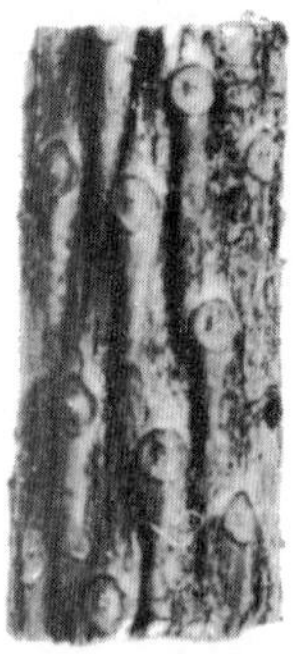

Both these Firs have strong balsam odor, needles dark green above, distinct whitish look underneath, the Fraser, especially, being almost silvery. Fraser Fir needles appear denser on the twig. Both feel soft compared with Spruces.

Notice round leaf scars. This is very characteristic of Firs, and can be best seen on older twigs where the needles have already fallen off.

Needles are flat, unlike four-sided needles of Spruce.

Buds are rounded, unlike sharp-pointed buds of Douglas Fir.

SCALE
ACTUAL SIZE

female
(pistillate)

Typical Fir flowers

male (staminate)

Fir cones grow upright on the branches. They mature in one season, then disintegrate leaving a spike.

Balsam Fir

1 ¼ — 2 ¾″

Fraser Fir

Notice "bracts" coming out between scales of cones. They are usually much more evident with the Fraser than with the Balsam Fir.

1 ¾ — 3 ¼″
average length

Balsam Fir

Bracts barely visible in left-hand cone. Compare with Fraser.

HEMLOCK—*Tsuga canadensis*
(Eastern or Canadian Hemlock)

Carolina Hemlock—*Tsuga caroliniana,* is a southern tree native to a very small area mostly in western N. Carolina. It is sometimes sold by nurseries as far north as New England.

Flowers Monoecious—May-June
Large
Found mostly in northern states and Appalachians

conelike female flowers at ends of twigs

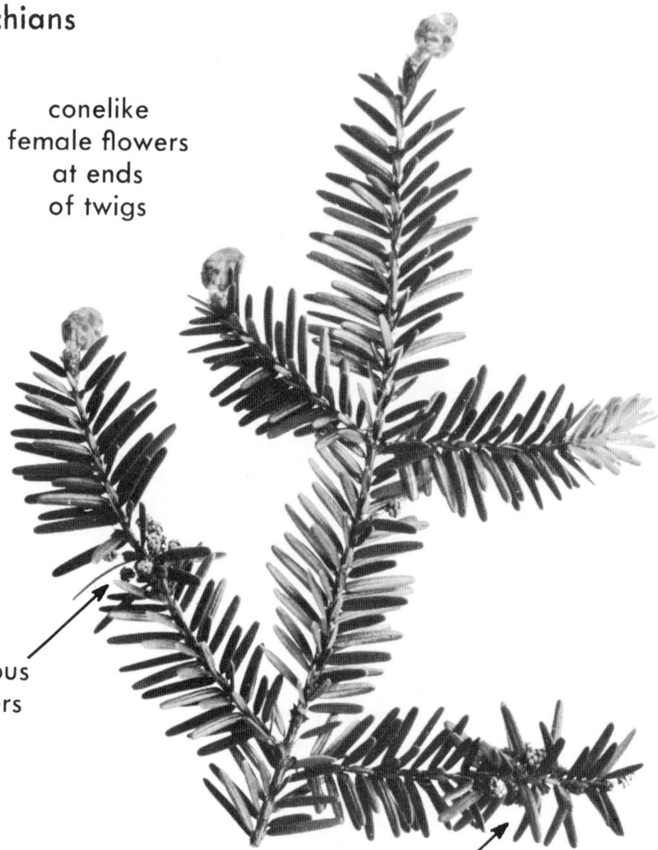

inconspicuous male flowers

male flowers

SCALE
ACTUAL SIZE

immature cones

cones mature in one season

The needles have two white (stomatic) bands underneath which give the whole underside a characteristic whitish look.

BALD CYPRESS—*Taxodium distichum* Deciduous (losing needles in fall)

Flowers Monoecious—March-April
Large
Found in southern part of range, along East Coast to Delaware and also extends up Mississippi River valley to Illinois. Very often grows in water, sending up strange growths from the roots, called "knees."

in winter

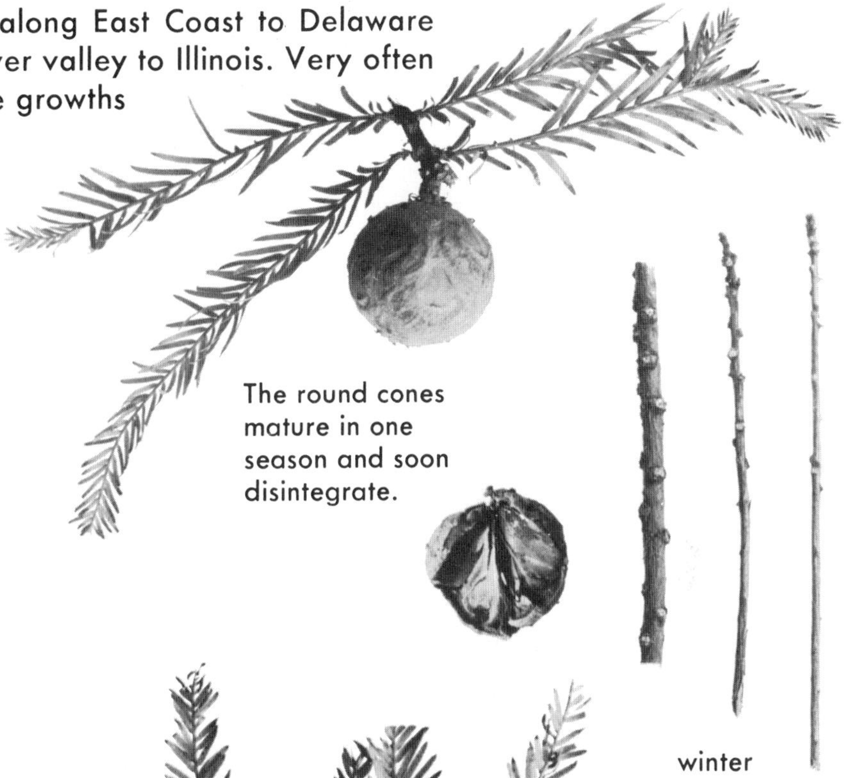

The round cones mature in one season and soon disintegrate.

winter twigs

Twigs, needles, fruit

SCALE
ACTUAL SIZE

in summer

distinctive knees

RED CEDAR—*Juniperus virginiana*

Flowers usually Dioecious, occasionally Monoecious—May
Small—medium
Found in most of range

Common Juniper *Juniperus Communis* (usually a shrub) always has prickly-type needles growing in threes around the stem.

The Red Cedar is sometimes broad and squat, at other times tall and narrow.

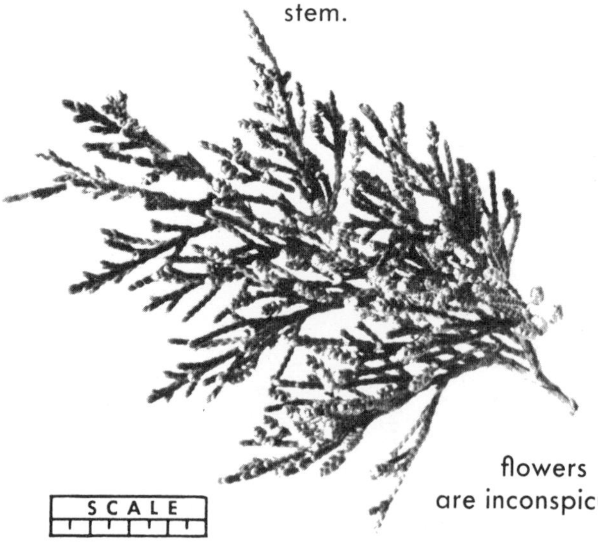

flowers are inconspicuous

SCALE
ACTUAL SIZE

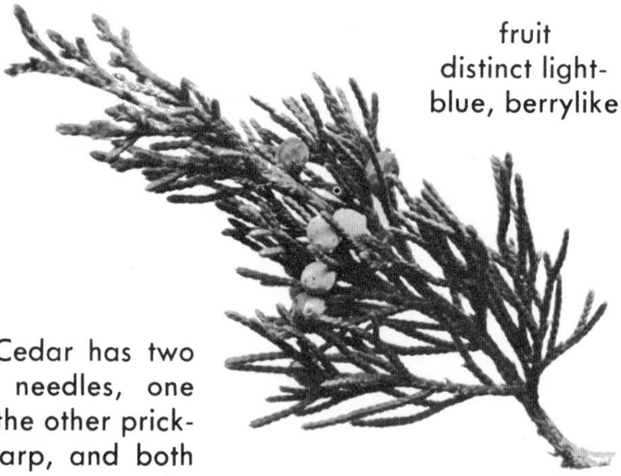

fruit distinct light-blue, berrylike

The Red Cedar has two types of needles, one scalelike, the other prickly and sharp, and both types are often on the same tree, although it almost never has the prickly-type only. Compare with White Cedar on opposite page.

This is a Cedar gall (Cedar apple). It is orange in color and often develops in wet weather, drying up in dry weather.

WHITE CEDAR—*Chamaecyparis thyoides*
(Atlantic White Cedar)

Flowers Monoecious—April-May
Medium to large
Found almost entirely in swamps, fairly near the coast from Maine to Texas

SCALE
ACTUAL SIZE

Note: The fruit of the White Cedar is a ball-like cone, quite different from the blue, berrylike fruit of the Red Cedar.

The needles of the White Cedar, which appear superficially like those of the Red Cedar, are really much flatter. Red Cedar needles feel four-sided and are obviously so when seen under a magnifying glass.

There are a large number of hybrids and varieties of *Chamaecyparis* sold by nurseries, usually under the name of Cypress. These are mostly of foreign origin. They vary in color from golden yellow to green. The green types might be mistaken for White Cedar, but the needles almost always have a whitish look underneath, unlike those of the native White Cedar, which are green on both sides.

ARBORVITAE—*Thuja occidentalis*
(Northern White Cedar)

This column
SCALE
ACTUAL SIZE

Flowers Monoecious—May
Medium
Mostly a northern tree, sometimes found in mountainous parts of South

Arborvitae has distinct cones, quite different from the Red Cedar's blue berries and White Cedar's round, indistinct cones.

The needles of Arborvitae are very flat, and easily distinguished from those of the Red Cedar and White Cedar, which are often confused with it, largely because of the popular names. Arborvitae is sometimes mistakenly called White Cedar. A more correct name is Northern White Cedar.

YEW – *Taxus*

Flowers usually Dioecious

There are no eastern Yews of tree size, the only member of the genus native to this part of the country being the Canadian Yew, *Taxus canadensis,* a low shrub. The Japanese Yew, in variety, is the one most commonly encountered, and the upright type, *Taxus cuspidata capitata,* is shown here. The English Yew, *Taxus bacata,* is also planted in this country, but more often as a cross with the Japanese Yew.

The Yew family is considered as distinct from the other evergreens which bear cones. The Yews all bear red berries.

This column
SCALE
ACTUAL SIZE

Upright Japanese Yew
(Taxus cuspidata capitata)

bright-
red berries

Yew needles are
dark green above,
yellow-green below

GINKGO – *Ginkgo biloba*
(Maidenhair Tree)

Flowers Dioecious—May
Medium to large
Brought to this country from China and Japan, now found over most of range

young leaves

flowers

This column
SCALE
ACTUAL SIZE

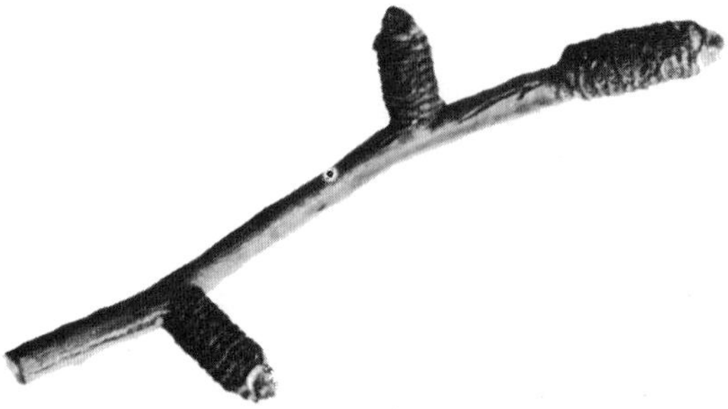

edible nut inside pulpy fruit

WILLOW – *Salix* Flowers Dioecious—April-May Easily propagated from cuttings

Black Willow	*Salix nigra*	medium	Native, found in most of range
Crack Willow	*S. fragilis*	large	Europe, N. Africa, Asia, now generally naturalized here
Weeping Willow (Babylon Weeping Willow)	*S. babylonica*	medium	Originally from Asia. There are a number of Weeping Willows, none are native
White Willow (Pussy Willows—a number of species, all shrubs)	*S. alba*	medium to large	Europe, W. Asia, has become naturalized here

Willow twigs are more colorful than those of most other trees, usually conspicuously so, which makes recognition easy in winter, even at a considerable distance. They often look bright yellow, red or olive-green.

Black Willow

White Willow

Crack Willow

Weeping Willow

Notice stipules around twig

SCALE
ACTUAL SIZE

Black Willow

Note: Sickle-shaped leaf. Stipules (leaflike appendages) persist around twig. Very dark, shiny leaf, green underneath and darker than other willows shown here. Buds thicker in relation to length.

White Willow

very white or silvery underneath

Crack Willow

Weeping Willow

Note: Willow buds are covered by a single scale, and hug closely to the stem. They start to grow early in the spring, the expanding flower buds becoming "Pussies." Poplars also do this.

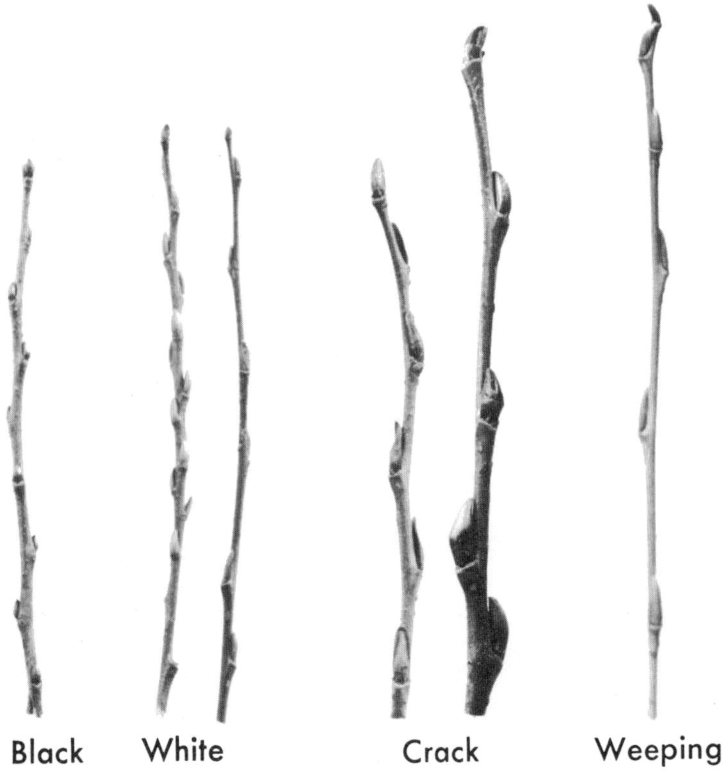

Black White Crack Weeping

Black Willow

flowers

SCALE
ACTUAL SIZE

fruit

Crack Willow

White Willow

POPLAR – *Populus*

Flowers Dioecious—April-May Easily propagated from cuttings

Balm of Gilead	*Populus candicans*	medium to large	Mostly northern states
Balsam Poplar (or Tacamahac)	*P. tacamahaca*	medium to large	Mostly northern states
Cottonwood	*P. deltoides*	medium to large	Most of range
Large-toothed Poplar (or Aspen)	*P. grandidentata*	usually medium	Large part of range
Lombardy Poplar	*P. nigra italica*	medium to large	Europe, most of range
Trembling (or Quaking) Aspen	*P. tremuloides*	medium to large	Large part of range
White Poplar	*P. alba*	medium to large	Europe, extensively here

Cottonwood

White Poplar

Trembling Aspen

Large-toothed Poplar

Balsam Poplar

Lombardy Poplar

bark on
old trees
becomes
furrowed

Large-toothed Poplar
(usually yellowish)

Trembling Aspen
(often very white, sometimes yellowish)

Balm of Gilead (Balsam Poplar similar)

Cottonwood

Note: Upper branches of many poplars are whitish. The black-spotted look is typical of White Poplar.

White Poplar

White Poplar

SCALE
ACTUAL SIZE

Balm of Gilead

Cottonwood

Balsam Poplar

Lombardy Poplar

SCALE
ACTUAL SIZE

immature leaves
of Large-toothed Poplar
very silvery for
brief period

Large-toothed Poplar

Trembling Aspen

White Poplar

The leaf stems of Trembling Aspen, Large-toothed Poplar and Cottonwood are very flat, those of the others less so

shiny, sticky buds

dull, coated buds

Lombardy | Trembling Aspen | Large-toothed | White Poplar | Balsam | Cottonwood | Balm of Gilead

SCALE
ACTUAL SIZE

Representative Poplar flowers in various stages of development

Trembling Aspen

"Pussy" beginning to open

Cottonwood

female (pistillate) flowers (notice development of these on opposite page)

Trembling Aspen

flower further developed

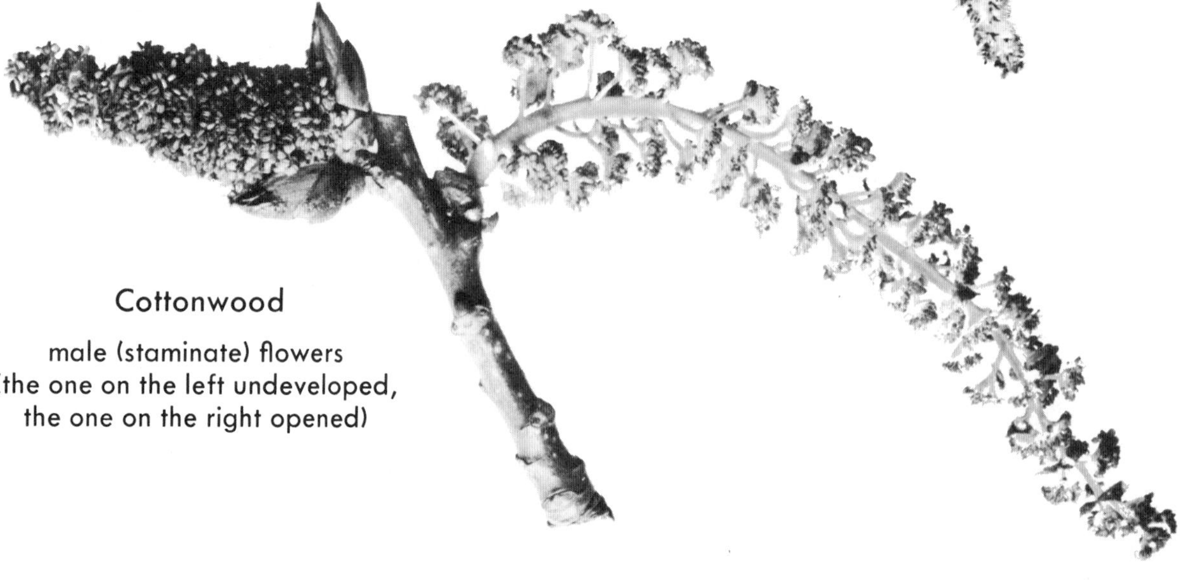

Cottonwood

male (staminate) flowers (the one on the left undeveloped, the one on the right opened)

SCALE
ACTUAL SIZE

Representative Poplar fruit

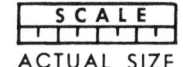

Large-toothed Poplar

note cotton

All the Poplars produce seeds attached to fluffy cotton, easily carried by the wind.

Poplars and Willows, both of the Willow family, are difficult to classify. They both have a great many species. The Poplars tend to cross among themselves, producing many confusing hybrids. This is also true of the Willows. Therefore, do not expect to find clear-cut identification always possible.

Cottonwood

note cotton

HICKORY – *Carya* Flowers Monoecious—May

Bitternut (Hickory)	*Carya cordiformis*	large	Most of range except extreme South
Mockernut (Hickory)	*C. tomentosa*	medium to large	Most of range except extreme North
Pecan	*C. pecan*	large	Mostly in Middle West, Illinois south
Pignut (Hickory)	*C. glabra*	large	Found mostly east of Mississippi River
Shagbark Hickory	*C. ovata*	large	Most of range except extreme North and South

Mockernut

Shagbark Hickory

Pignut

Bitternut

Pecan

SCALE

1/3 ACTUAL SIZE

Pecan

Bitternut

Average number of leaflets
Pecan — 9-17
Bitternut — 5-11
Mockernut — 5-9
Pignut — 5-7, usually 5
Shagbark — 5-7, usually 5

Shagbark
(Pignut and Mockernut leaves look similar)

The size, shape and number of leaflets of these compound Hickory leaves are important details to notice.

The Pecan and Bitternut have somewhat similar leaves, but the Pecan has the *largest* leaf of the Hickories, Bitternut the *smallest*. Notice the sickle-shaped leaflet on the Pecan (top left Pecan leaflet). This is typical of Pecan.

The Shagbark, Pignut and Mockernut have three large leaflets at the end of the leaf. The Mockernut leaf is particularly fragrant when crushed.

typical Hickory
flowers

SCALE
ACTUAL SIZE

Pecan

Distinctive yellow buds
Pecan
Bitternut

Pignut

Very bitter taste
Bitternut

Thin husks
Pecan
Bitternut
Pignut

Bitternut

Mockernut

Thick husks
Mockernut
Shagbark

Shagbark

Shagbark Hickory

Mockernut

Pignut

Bitternut

Pecan

WALNUT—*Juglans* Flowers Monoecious—May

Black Walnut	*Juglans nigra*	large	Most of range
Butternut (White Walnut)	*J. cinerea*	medium	Most of range except deep South

Black Walnut

Butternut

Black Walnut
(dark bark)

Butternut
(light bark)

SCALE
ACTUAL SIZE

Typical Walnut flowers

Butternut
Twigs have dark brown, finely chambered pith

Black Walnut
Twigs have light brown, coarsely chambered pith

Black Walnut
(usual number of leaflets 15-23)

SCALE
½ ACTUAL SIZE

Butternut
(usual number of leaflets 11-19)

HACKBERRY—*Celtis occidentalis*

Flowers Monoecious—May
Medium to large
Found in most of range except extreme South

Leaves, flowers, fruit, twigs

SCALE
ACTUAL SIZE

rough-type bark

smooth-type bark

MP 41

ELM – *Ulmus* Flowers Perfect or Monoecious—April

American Elm	*Ulmus americana*	large	Most of range
Slippery Elm	*U. fulva*	medium to large	Most of range except not often found along coast south of Delaware

American Elm

Slippery Elm

very fuzzy buds

Slippery Elm

smooth, shiny buds

American Elm

SCALE

ACTUAL SIZE

typical Elm flowers

American Elm

notch in end of wing

no notch in wing; larger than American Elm

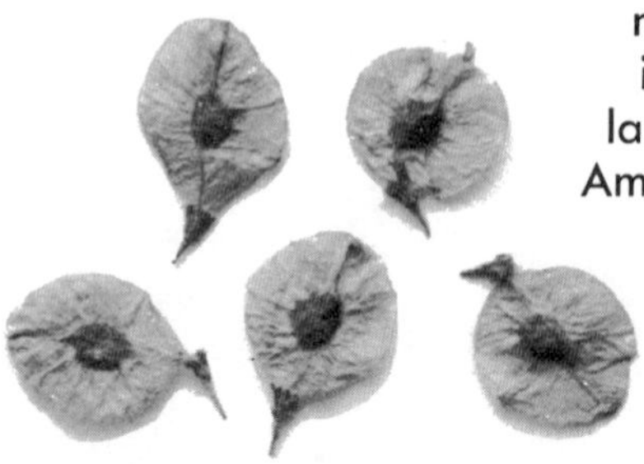

Slippery Elm

American Elm

Note: Upper side of Slippery Elm feels rough in *all* directions. Smaller teeth than American Elm.

SCALE

ACTUAL SIZE

Slippery Elm

American Elm (smooth-type bark)

American Elm (furrowed bark)

Slippery Elm

BIRCH — *Betula* Flowers usually Monoecious, rarely Dioecious—April-May

Black (Sweet or Cherry) Birch	*Betula lenta*	medium to large	Mostly New England Appalachians
Gray Birch	*B. populifolia*	small to medium	Mostly northeast
Red (or River) Birch	*B. nigra*	medium to large	South to S. New England
Paper (White or Canoe) Birch	*B. papyrifera*	medium to large	Northern tree
Yellow Birch	*B. lutea*	medium to large	North and Appalachians

Paper Birch

Gray Birch

Yellow Birch

Black Birch

Red Birch

SCALE
ACTUAL SIZE

Gray and Red Birches have smaller catkins, buds and twigs than the others.

Gray and Red Birches

Black and Yellow Birches

The twigs of these two are very similar, and both have characteristic sweet "Birch" taste and smell.

Notice winter catkins typical of Birches, (also Hop Hornbeam and Alders)

Paper Birch

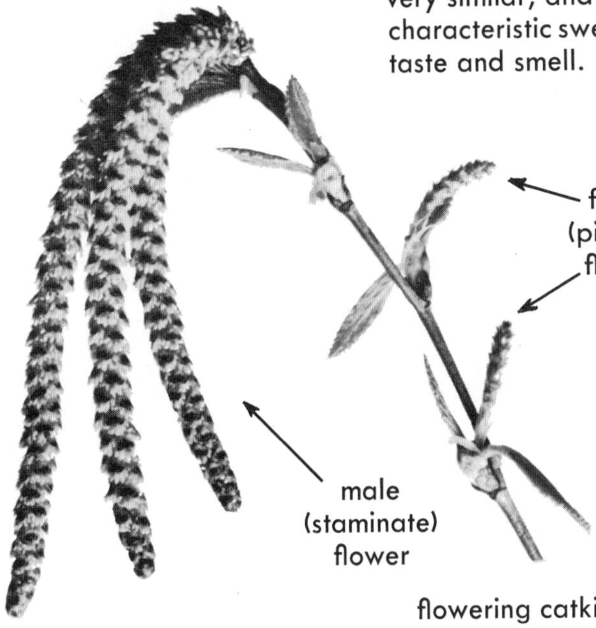

flowering catkins typical of birches

fruiting catkins

Gray Birch

Paper Birch

Black, Yellow and Red Birches have similar fruit.

Gray Birch

Paper Birch

Black Birch (young bark)

Black Birch (old bark)

Red Birch (young bark)

Red Birch (old bark)

Leaves
SCALE
ACTUAL SIZE

Gray Birch

Paper Birch

Black Birch

Yellow Birch

Red Birch

Yellow Birch (old bark)

Yellow Birch (young bark)

AMERICAN BEECH – *Fagus grandifolia*

This column
SCALE
ACTUAL SIZE

Flowers Monoecious—May
Large
Most of range

flowers

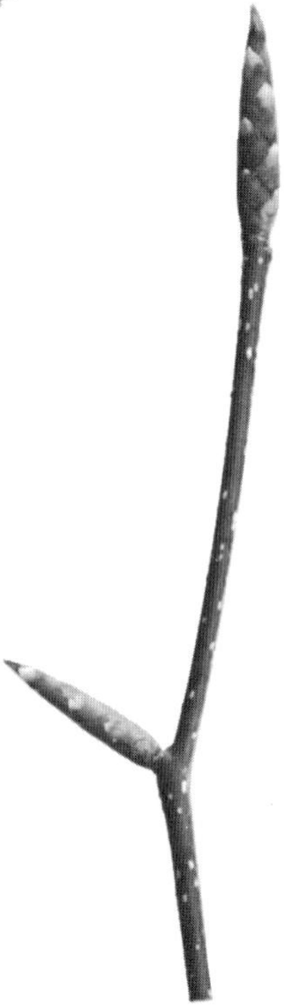

Beech nuts and burrs

typical smooth bark

MP
48

AMERICAN CHESTNUT – *Castanea dentata*

Flowers Monoecious—July
Large (see note below)
Found mostly east of Mississippi River

Note: The American Chestnut has disappeared as a large tree since the Chestnut blight was brought from Asia about the beginning of the twentieth century. It can still be found coming up from old roots, and occasionally produces fruit before dying of the blight.

Although the Chinese and European Chestnuts are blight resistant, they are inferior as timber to our native tree, and no successful hybrid has been found.

Smooth bark when young. Notice cracked, diseased bark at base, symptomatic of the blight.

female
(pistallate)
flowers

This column
SCALE
ACTUAL SIZE

male
(staminate)
flower

AMERICAN HORNBEAM – *Carpinus caroliniana*
(Blue Beech or Ironwood)

This column
SCALE
ACTUAL SIZE

Flowers Monoecious—April-May
Small
Most of range

flowering
catkins

Typical muscular-looking trunks
and smooth bark

winged
seeds

AMERICAN HOP HORNBEAM – *Ostrya virginiana*
(Ironwood)

This column
SCALE
ACTUAL SIZE

Flowers Monoecious—April-May
Medium
Most of range

distinctive winter catkins — often in threes — which develop into flowering catkins.

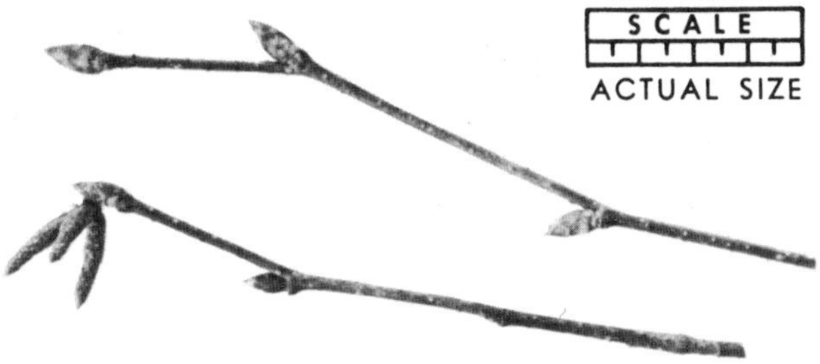

distinctive fruit and bark

old bark

young bark

OAKS – *Quercus* Flowers Monoecious—May

WHITE OAK GROUP—Acorns mature in one year – Lobes of leaves rounded

Burr Oak	*Quercus macrocarpa*	large	Most of range except southeastern states
Chestnut Oak	*Q. montana*	medium to large	Eastern half of range except not along coast south of N. J.
Swamp White Oak	*Q. bicolor*	medium to large	Mostly northern tree
White Oak	*Q. alba*	large	Most of range

White Oak

Swamp White Oak

Chestnut Oak

Burr Oak

RED (or BLACK) OAK GROUP—Acorns mature in two years
Lobes of leaves sharp and bristle-pointed

Black Oak	*Quercus velutina*	large	Most of range
Pin Oak	*Q. palustris*	large	Most of range, not extreme North or South
Red Oak	*Q. borealis*	large	Most of range except extreme South
Scarlet Oak	*Q. coccinea*	large	Mostly east of Mississippi and not much in extreme North or South
Willow Oak	*Q. phellos*	large	Mostly found in southern part of range, but hardy except extreme North

Red Oak

Black Oak

Pin Oak

Scarlet Oak

Willow Oak

SCALE
ACTUAL SIZE

typical Oak flowers
both White and Red (or Black)
Oak groups

acorns and twigs
of White Oak group

Note: All Oaks tend to have a cluster
of buds at the ends of the twigs.

Burr Oak

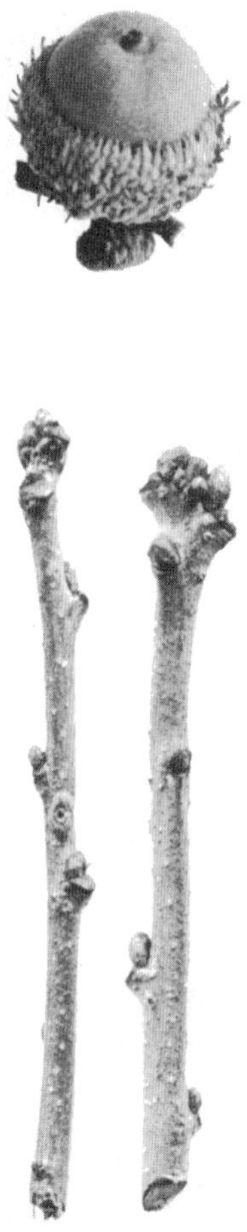

Swamp White Oak

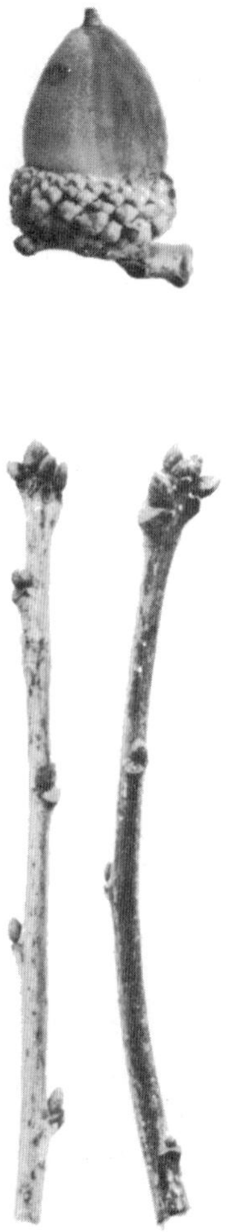

White Oak

Chestnut Oak

acorns and twigs
of Red (or Black) Oak group

SCALE
ACTUAL SIZE

notice
very shallow cup,
large acorn

Red Oak

Notice deep cup, scales of which are fairly smooth and *do not stick out* — acorn similar to Black Oak

Scarlet Oak

Willow Oak

notice deep cup and
scales which *tend
to stick out*

Black Oak

Pin Oak

top half of buds fuzzy,
bottom half shiny

Red Oak

fuzzy,
dull tan buds

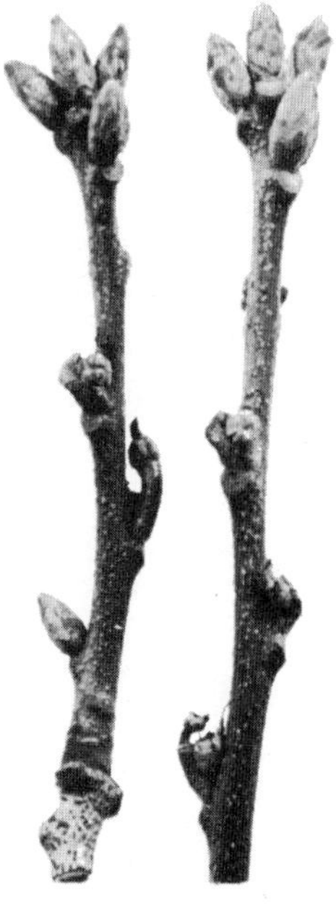

Black Oak

top half of buds fuzzy,
bottom half shiny

Scarlet Oak

Pin Oak

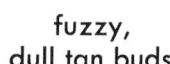

Willow Oak

White Oak Group

White Oak

White Oak

White Oak

Swamp White Oak

Chestnut Oak

Burr Oak

Red (or Black) Oak Group

Scarlet Oak

Red Oak

Black Oak

Pin Oak

probable hybrid—Black and Scarlet Oak

Willow Oak

SCALE
½ ACTUAL SIZE

White Oak Group (rounded lobes)

White Oak

large middle sinuses

Burr Oak

Chestnut Oak

Swamp White Oak

The White Oak group is readily distinguished from the Red Oak group by its leaves. Within the same group, however, leaves alone must not be used for final identification, as, even on the same tree, the leaves may vary more among themselves than between those of other species. This does not mean that there is no typical shape or shapes for any given species, but only that not every leaf will take a typical form. Therefore, leaves are of great assistance if: (1) an average leaf is looked for, and (2) final identification is based on other details taken in conjunction with the leaves.

Red (or Black) Oak Group

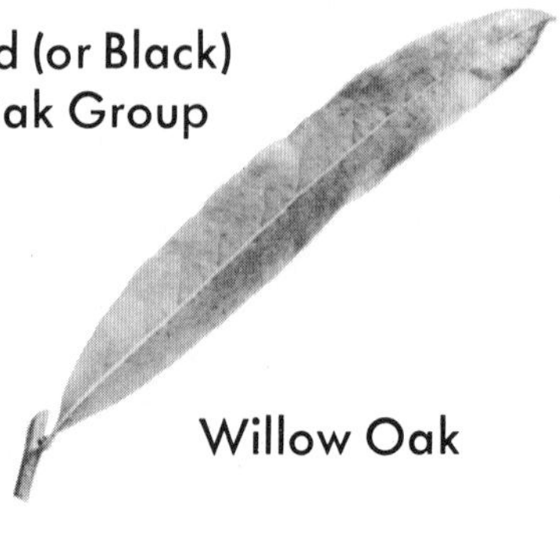

Willow Oak

Red (or Black) Oak Group (sharp, bristle-pointed lobes)

SCALE
½ ACTUAL SIZE

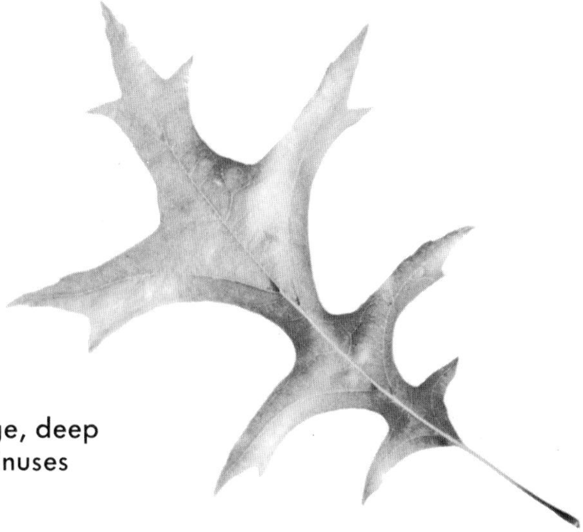

large, deep sinuses

Pin Oak

Scarlet Oak

Black Oak

Red Oak

MULBERRY – *Morus* Flowers Monoecious or Dioecious—May-June

Red Mulberry	*Morus rubra*	medium	Most of range except extreme North
White Mulberry	*M. alba*	medium	From China—extensively planted eastern U. S.

Note: There are a number of foreign Mulberries, or varieties of the White Mulberry, with red or black fruit. However, the leaves are usually shiny and much like White Mulberry leaves, distinguishing them from the native Red Mulberry.

Mulberry

typical Mulberry bark

typical Mulberry female (pistillate) flowers

This column SCALE ACTUAL SIZE

Red Mulberry

red to black fruit

white fruit

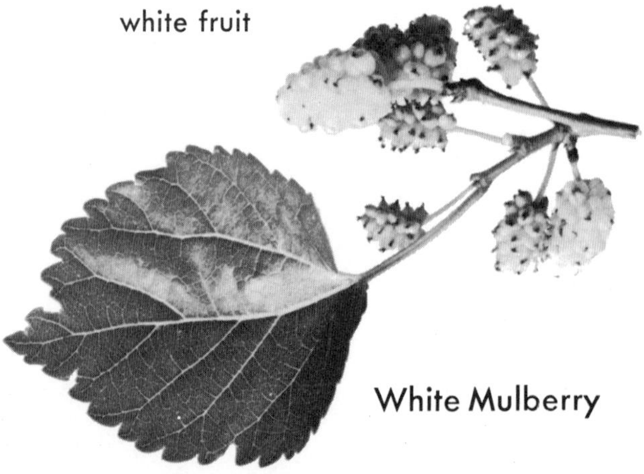

White Mulberry

SCALE

ACTUAL SIZE

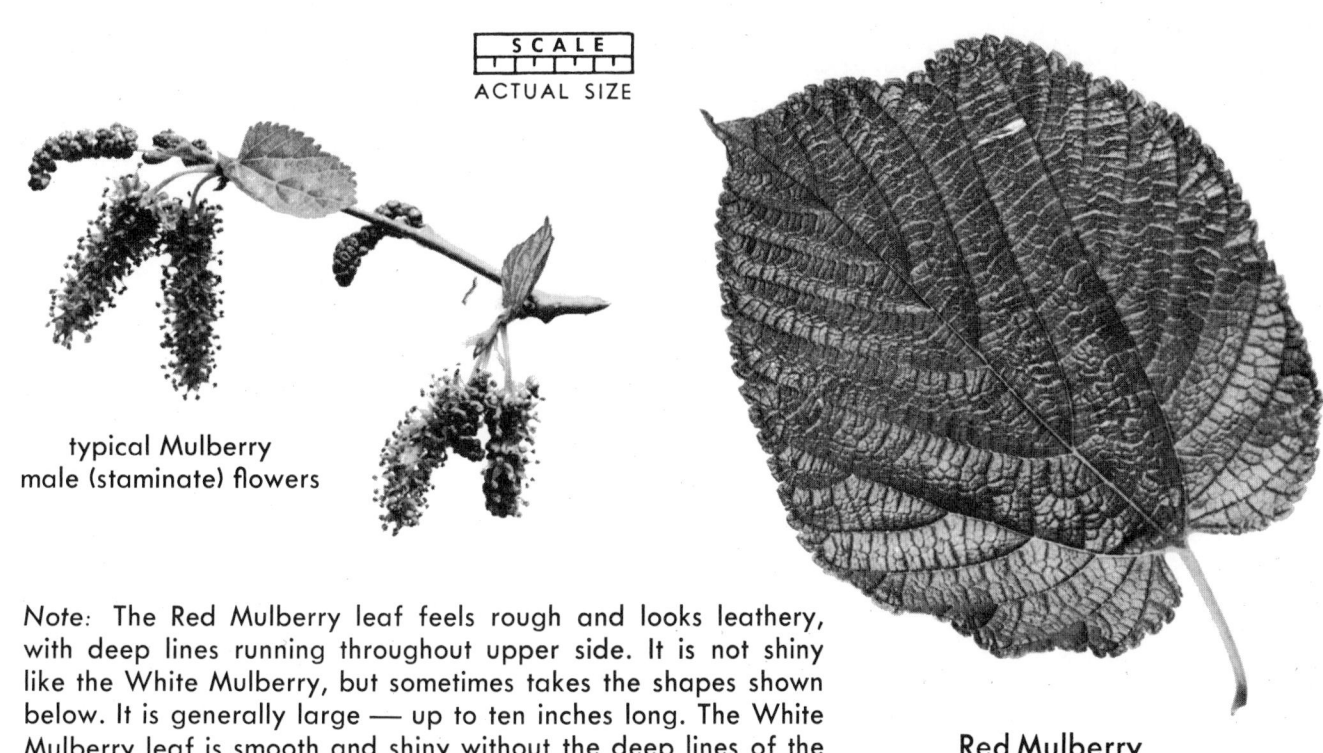

typical Mulberry
male (staminate) flowers

Red Mulberry

Note: The Red Mulberry leaf feels rough and looks leathery, with deep lines running throughout upper side. It is not shiny like the White Mulberry, but sometimes takes the shapes shown below. It is generally large — up to ten inches long. The White Mulberry leaf is smooth and shiny without the deep lines of the Red Mulberry.

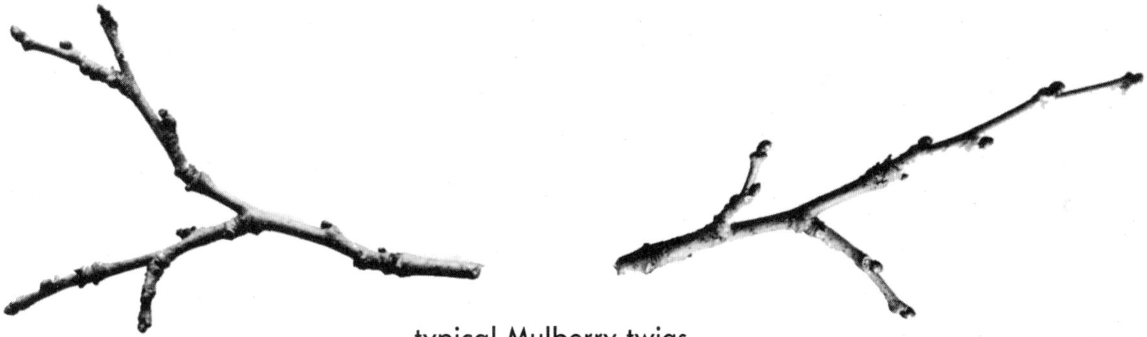

typical Mulberry twigs

White Mulberry
three shapes typical of Mulberries

OSAGE ORANGE – *Maclura pomifera*

Flowers Dioecious—May-June
Medium
Originally over small section of Texas,
Oklahoma and Arkansas—now
planted in most of range

This column
SCALE
ACTUAL SIZE

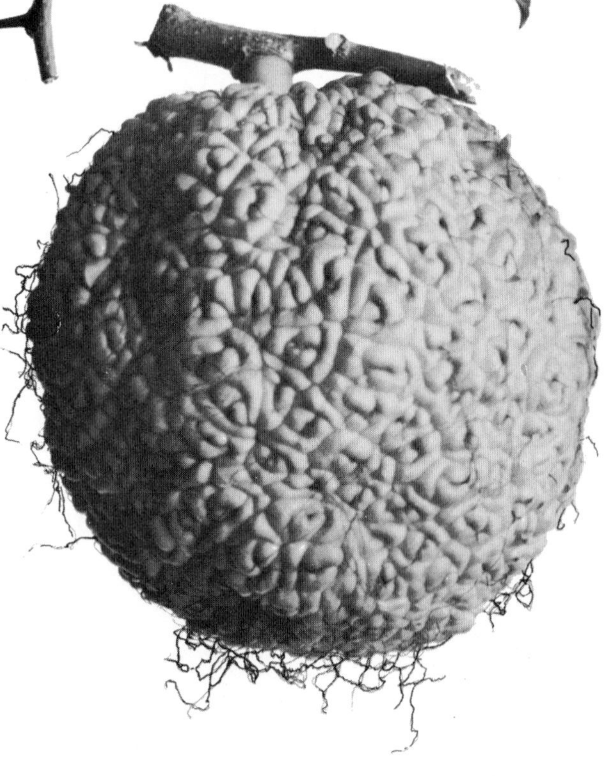

PAWPAW – *Asimina triloba*

Flowers Perfect—May
Small to medium
Southern, not found much
north of New Jersey

This column
SCALE
ACTUAL SIZE

distinctive
smell when
crushed

edible
fruit

MAGNOLIA – *Magnolia* Flowers usually Perfect—May-June

Cucumber Tree	*Magnolia acuminata*	medium to large	Hardy over much of range
Sweet Bay (or Swamp Magnolia)	*M. virginiana*	medium (small in North)	Hardy to Mass.
Umbrella Tree	*M. tripetala*	small to medium	Hardy over much of range

Cucumber Tree

Umbrella Tree

Cucumber Tree

Umbrella Tree

Sweet Bay
(evergreen in the South)

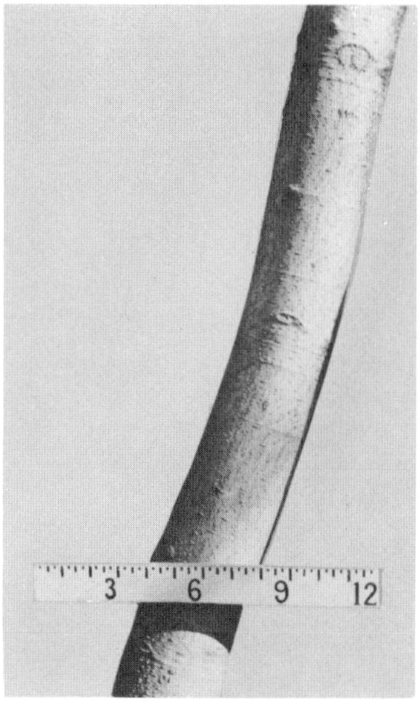

Sweet Bay

SCALE
ACTUAL SIZE

Sweet Bay

Umbrella Tree

SCALE
½ ACTUAL SIZE

Cucumber Tree

SCALE
½ ACTUAL SIZE

(See Flower Key, 15, for actual size picture)

Cucumber Tree

Umbrella Tree

SCALE
ACTUAL SIZE

Cucumber Tree

Umbrella Tree

Sweet Bay

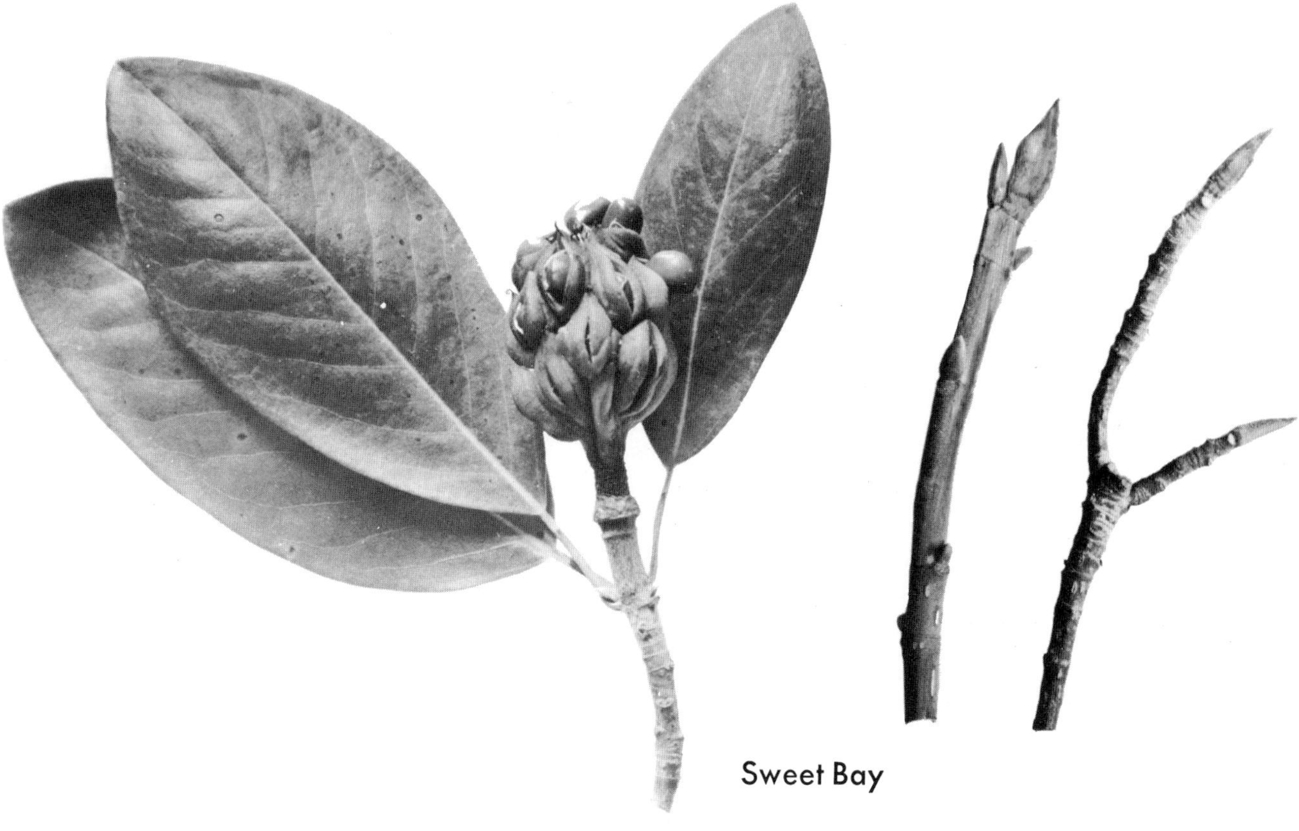

Sweet Bay

TULIP TREE — *Liriodendron tulipifera*
(White Wood or Yellow Poplar)

Flowers Perfect—June
Large
Mostly east of Mississippi River

notice unique leaf without a point at the end

The Tulip Tree has spreading branches when growing in the open, but in the woods it stands out from other trees by its straight, towering trunk with branches at the very top.

distinctive buds

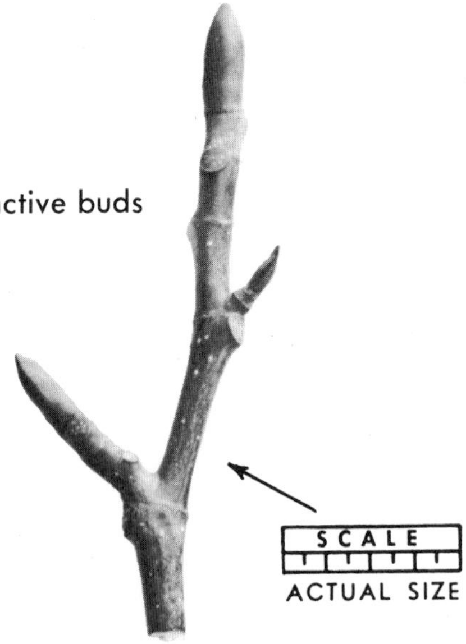

Flower and fruit
SCALE
ACTUAL SIZE

bark of young tree

immature and mature fruit

winged seed

SASSAFRAS – *Sassafras albidum*

Flowers Dioecious (occasionally Perfect)—May
Medium to large
Most of range except extreme North.

these three types of leaves often appear on same tree

Leaves
SCALE
½ ACTUAL SIZE

aromatic, smooth, green twigs

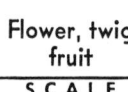

Flower, twig fruit
SCALE
ACTUAL SIZE

dark-blue berries in bright-red cups on red stems

SWEET GUM—*Liquidambar styraciflua*
(Bilsted)

Flowers usually Monoecious—April-May
Large
Found in much of range except extreme
North, especially common in South

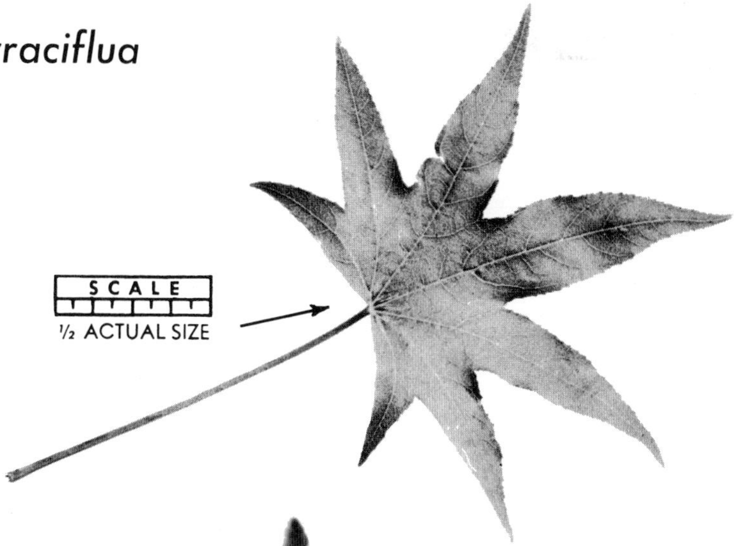

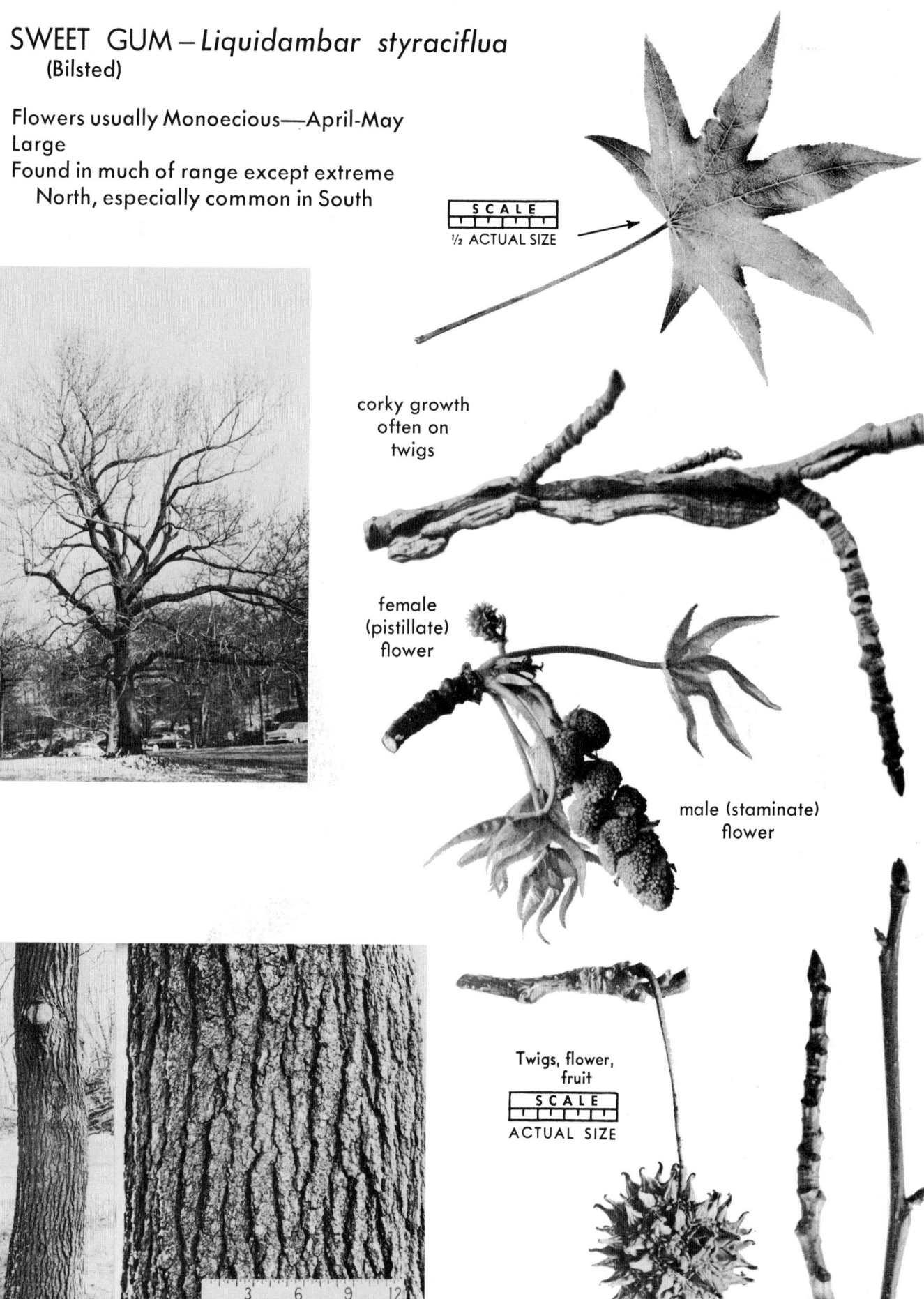

SCALE
½ ACTUAL SIZE

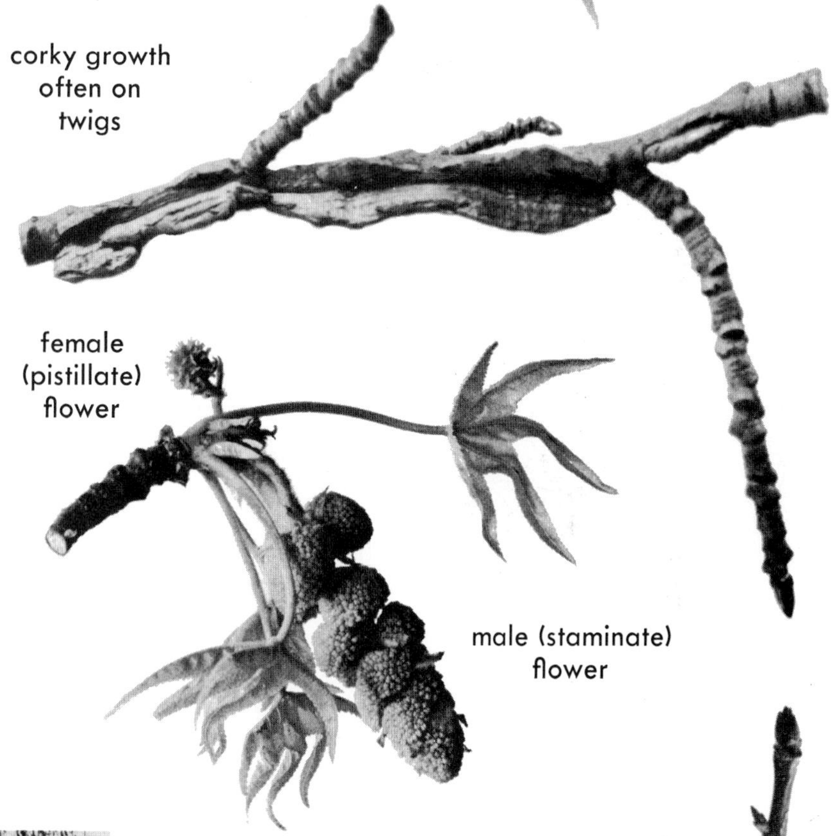

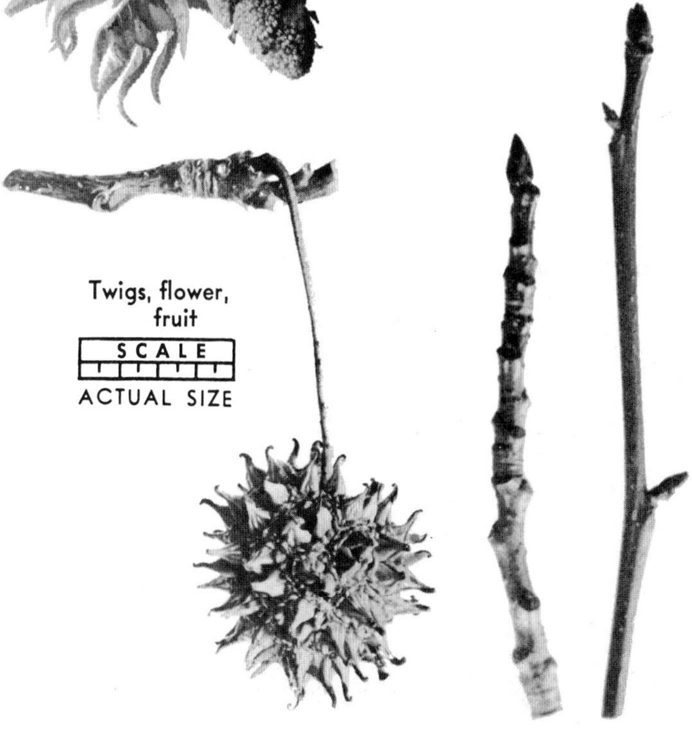

SYCAMORE – *Platanus occidentalis*

(Buttonball, Buttonwood or American Plane Tree)

Flowers Monoecious—May
Large
Found in most of range except extreme North

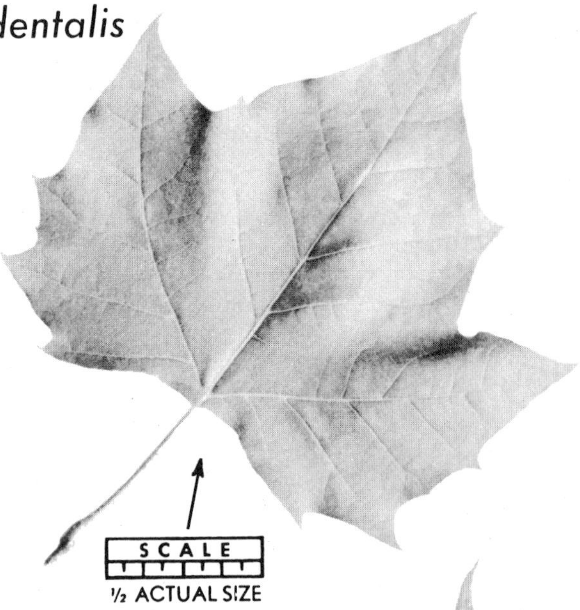

Flowers, twig fruit
SCALE
ACTUAL SIZE

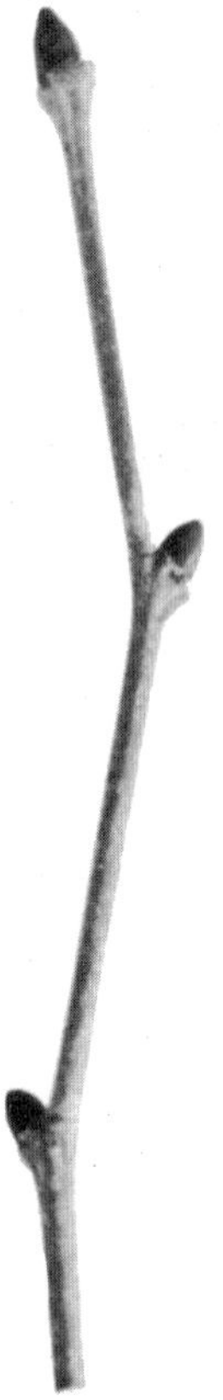

mature fruit
(hairy when young
—see Fruit Key)

HAWTHORN – *Crataegus*

This column
SCALE
ACTUAL SIZE

Flowers Perfect—May
Small
Most of range—800-1000 species in North America alone

notice different leaf types

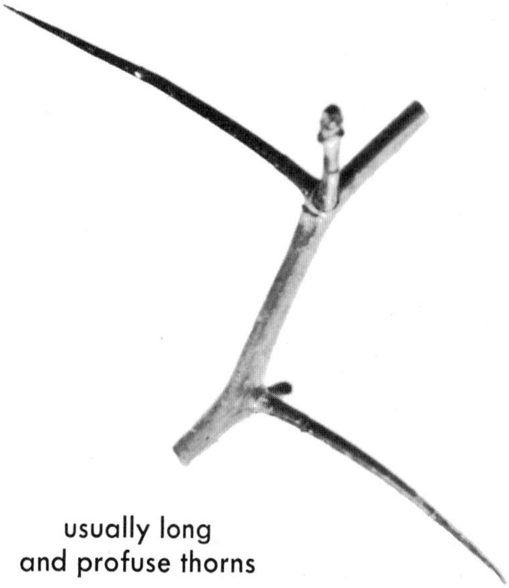

usually long and profuse thorns

MOUNTAIN ASH—*Sorbus* Flowers Perfect—May-June

American Mountain Ash	*Sorbus americana*	small	Prefers cold, mountainous country of northern states, but is found South in Appalachians
European Mountain Ash (Rowan Tree)	*S. aucuparia*	small to medium	From Europe. Extensively planted in U.S. Will grow farther south than American Mountain Ash

American Mountain Ash

smooth, nonhairy buds

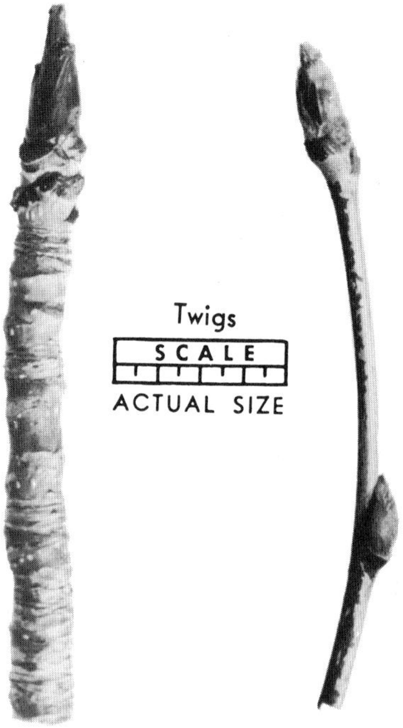

Twigs

fuzzy buds

European Mountain Ash

American Mountain Ash

European Mountain Ash

American Mountain Ash

European Mountain Ash

Leaves
SCALE
½ ACTUAL SIZE

CHERRY – *Prunus*   Flowers Perfect—May-June

Black Cherry (Rum Cherry)	*Prunus serotina*	medium to large	Most of range
Choke Cherry	*P. virginiana*	small, mostly shrub-like, often confused with Black Cherry	Most of range
Pin Cherry (Wild Red, Bird, or Fire Cherry)	*P. pennsylvanica*	small	Mostly northern, but found South in Appalachians
Sweet (or Mazzard) Cherry	*P. avium*	medium to large	From Europe, Asia, now much of range

Black Cherry

Pin Cherry

Sweet Cherry

Black Cherry

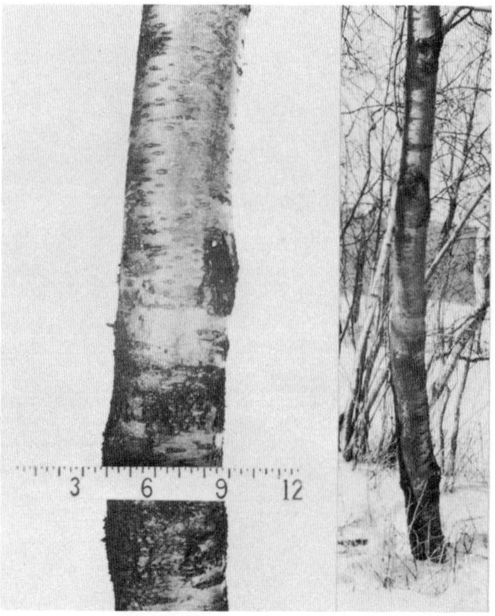

Pin Cherry (young bark shiny red)

Leaves, twigs

SCALE

ACTUAL SIZE

Sweet Cherry

Black Cherry

Choke Cherry

Pin Cherry

Sweet Cherry

Sweet Cherry

tent caterpillar egg mass often found on Cherry trees

Black Cherry
(Choke Cherry similar, both have unpleasant taste and smell)

Pin Cherry
very small buds

MF 77

SCALE
ACTUAL SIZE

Black Cherry
(Choke Cherry similar but blooms a week to ten days sooner than Black Cherry)

Pin Cherry

Sweet Cherry

Pin Cherry

Sweet Cherry

Black Cherry
(Choke Cherry similar in looks, but very sour in taste. Black Cherry sweet when ripe)

SHADBUSH – *Amelanchier canadensis*
(Juneberry or Serviceberry)

Flowers Perfect—April-May
Small, occasionally medium
Most of range

distinctive lines
in bark

Leaves, flowers,
fruit, twigs
SCALE
ACTUAL SIZE

This column
SCALE
ACTUAL SIZE

AMERICAN REDBUD – *Cercis canadensis*

(The European Redbud, *C. siliquastrum* is called The Judas Tree)

Flowers Perfect—April-May
Small
Most of range except not in northernmost states

Note: The flowers are borne along the larger twigs and small branches and not out at the ends as with most trees. This is a distinctive feature, and during much of the year flower buds and usually pods can be seen all along the small branches, whereas the ends of the twigs produce mostly indistinct little growth buds.

BLACK LOCUST—*Robinia pseudoacacia*
(Common or Yellow Locust)

This column
SCALE
ACTUAL SIZE

Flowers Perfect—June
Medium to large
Originally mostly in
Appalachians, now
found in most of range

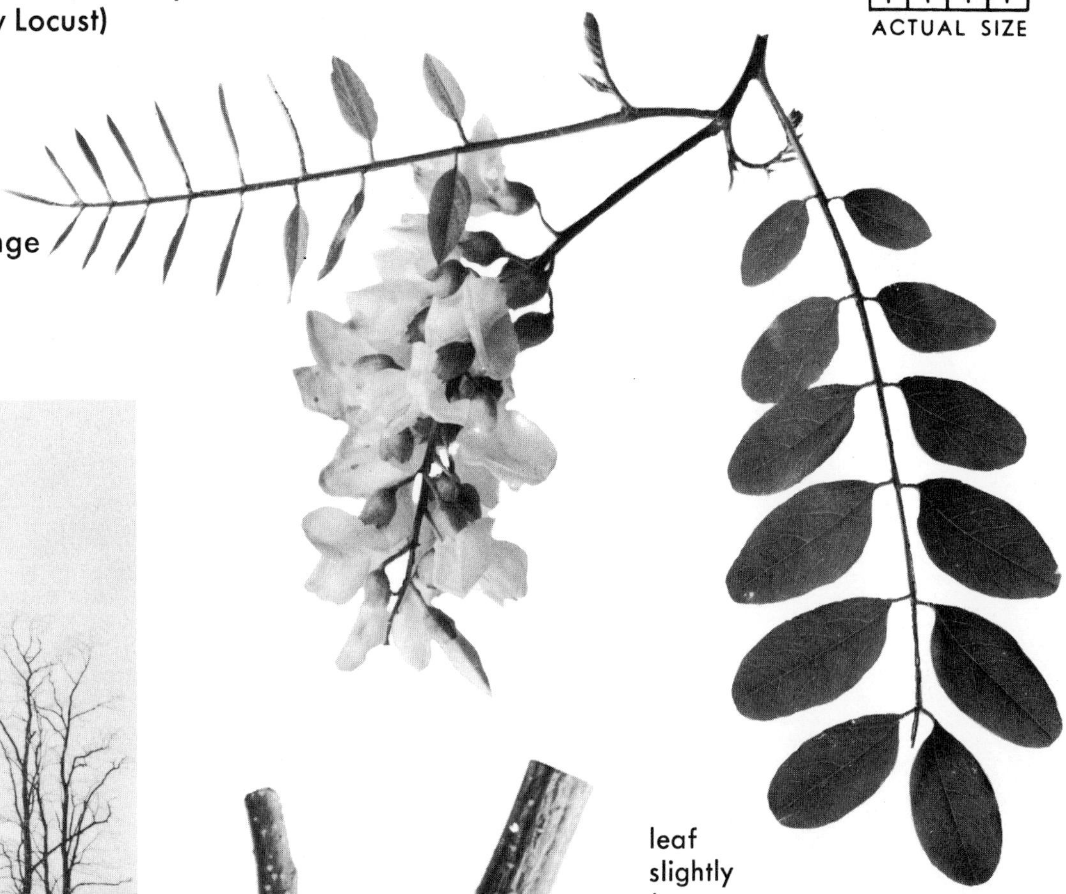

leaf
slightly
immature —
mature leaf
up to one
foot long

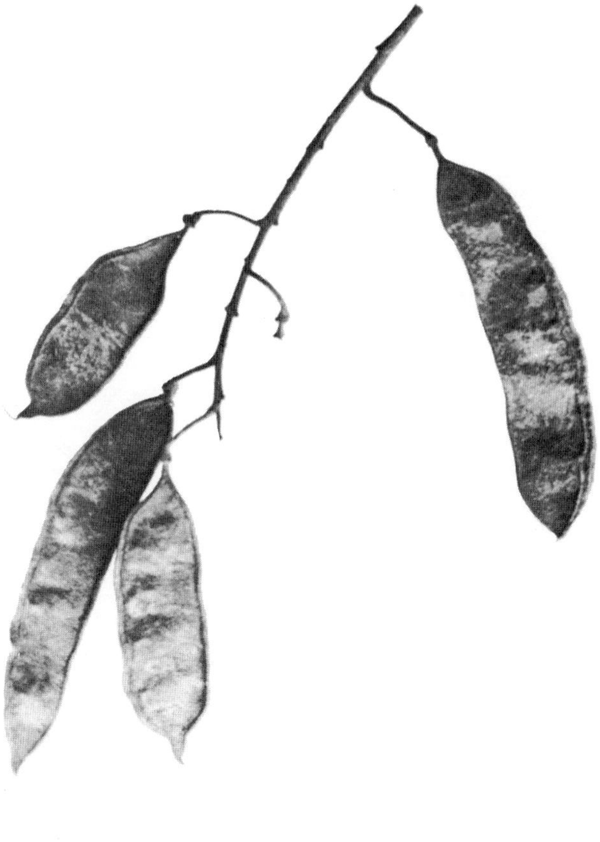

MP
81

HONEY LOCUST—*Gleditsia triacanthus*

(The Swamp or Water Locust, *G. aquatica,* is a southern tree found in swamps. It is a medium-sized tree and is very similar to the Honey Locust except for its pods, shown on opposite page.)

Flowers Monoecious (sometimes also Perfect)—June
Large
Found in much of range

inconspicuous
flowers

Flower, twig
SCALE
ACTUAL SIZE

notice thorns on
trunk and branches

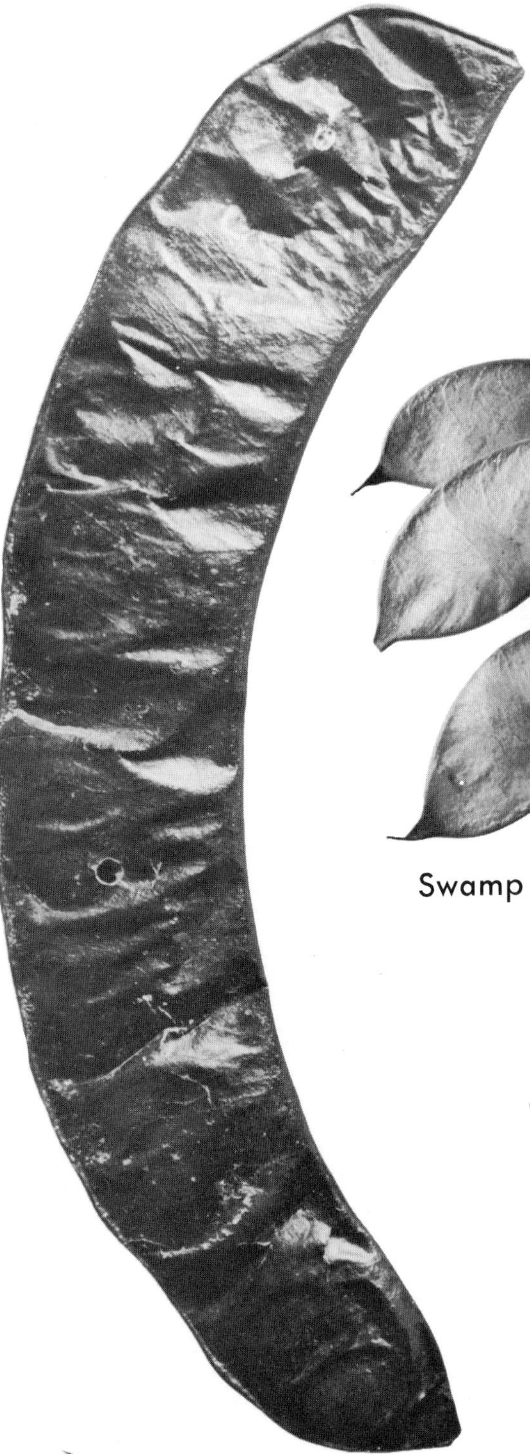

Swamp Locust

compound leaf,
occasionally doubly
compound, as below

KENTUCKY COFFEE TREE – *Gymnocladus dioicus*

Flower, pod, twig
SCALE
ACTUAL SIZE

Flowers Dioecious—June
Large
Originally in central part of range; now found occasionally over much of it

upper trunk and branches have very odd, shaggy bark

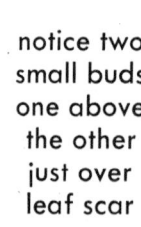

notice two small buds one above the other just over leaf scar

SCALE

½ ACTUAL SIZE

these leaves
are sometimes
three feet long

YELLOWWOOD — *Cladrastis lutea*

Flowers Perfect—June
Medium
Originally in very limited area, mostly parts of Tennessee and neighboring states, now found in most of range

SCALE
½ ACTUAL SIZE

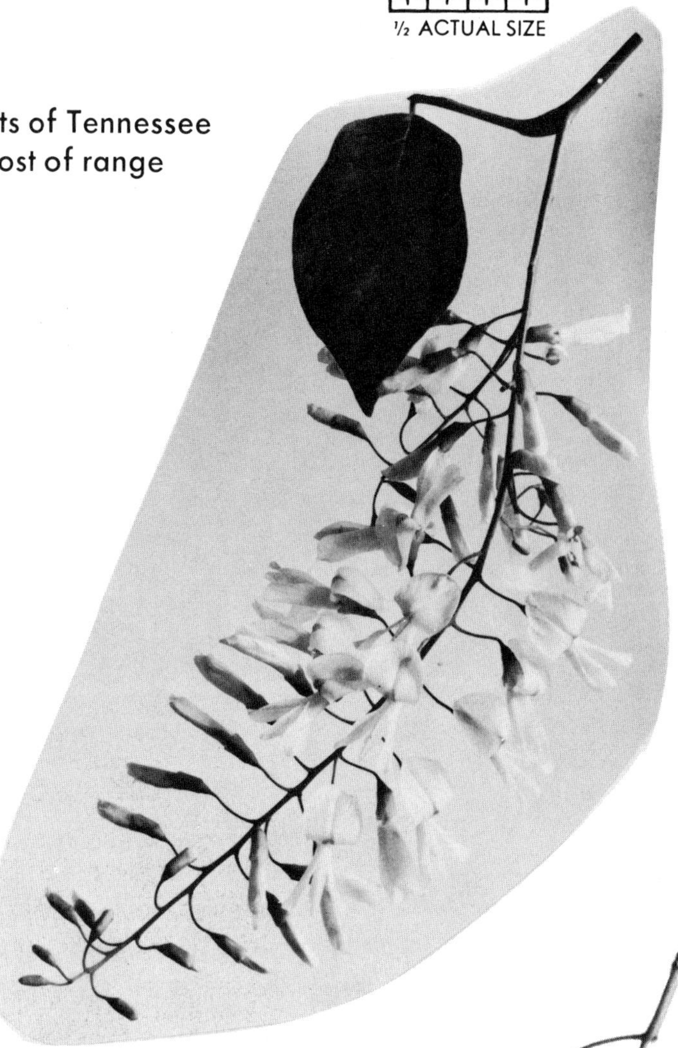

SCALE
ACTUAL SIZE

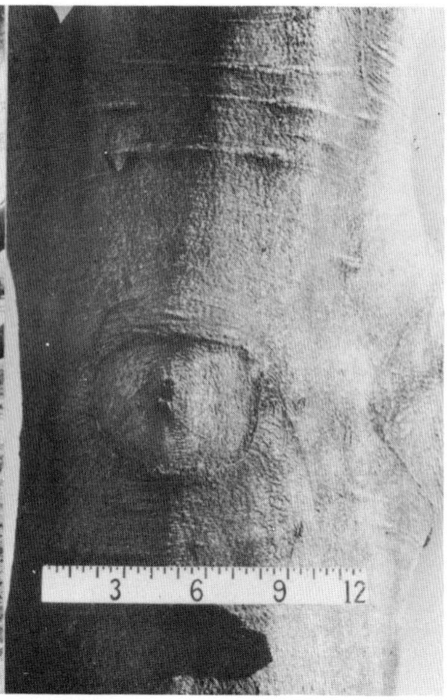

SCALE

½ ACTUAL SIZE

SCALE

ACTUAL SIZE

Note: The leaf stem completely covers the bud and is often found adhering to the twig after the individual leaflets have fallen. The leaf stem of Sycamore, also, completely covers the bud. Buds on most trees form on the twig just above the leaf, and therefore, after the leaf has fallen, the bud is seen above the leaf scar (the place where the leaf stem breaks off the twig). The leaf scars of Yellowwood and Sycamore, however, completely surround the buds.

HOP TREE – *Ptelea trifoliata*
(Wafer Ash)

This column
SCALE
ACTUAL SIZE

Flowers Perfect or Monoecious—June
Small
Much of range

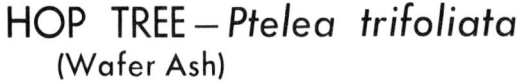

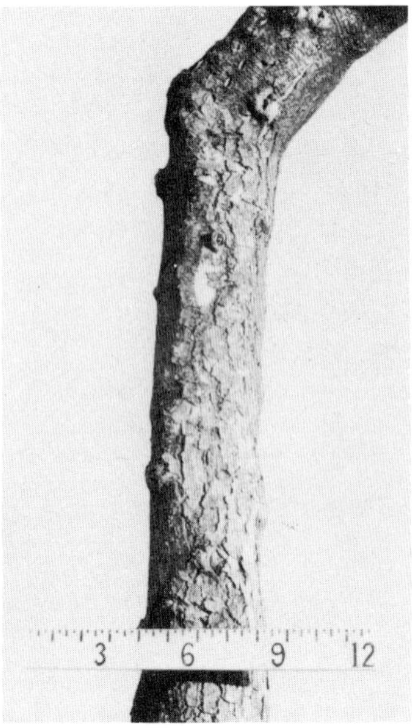

old fruit stems

MP
88

AILANTHUS – *Ailanthus altissima*
(Tree of Heaven)

Flowers usually Dioecious—June
Medium to large
From Asia, now over much of range

Flower, leaf
SCALE
1/2 ACTUAL SIZE

the leaf varies greatly in length — up to three feet with twenty-five or more leaflets

Fruit, twig
SCALE
ACTUAL SIZE

SUMACS – *Rhus*   Flowers Dioecious—June-August (see individual flowers)

Dwarf Sumac	*Rhus copallina*	small	Most of range
Poison Sumac	*R. vernix*	small	Most of range except extreme North. Almost always in swamps.
Smooth Sumac	*R. glabra*	small (smallest)	Most of range
Staghorn Sumac (Velvet Sumac)	*R. typhina*	small (largest)	Most of range

Staghorn Sumacs (typical of Sumac growth)

SCALE
2/3 ACTUAL SIZE

Poison Sumac
(blooms June)

Twigs
SCALE
ACTUAL SIZE

very fuzzy
twigs

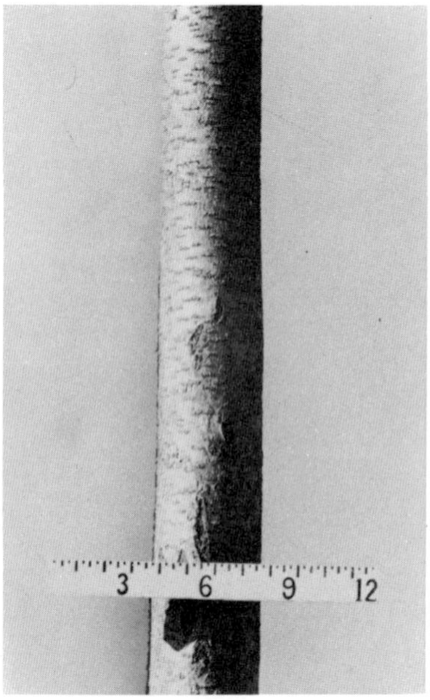

Staghorn Sumac
(typical Sumac bark)

Staghorn
Sumac

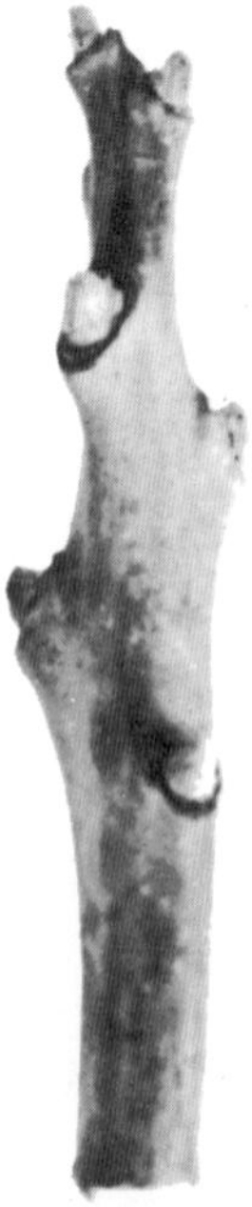

Smooth
Sumac

Dwarf
Sumac

Poison
Sumac

SCALE

⅔ ACTUAL SIZE

Dwarf Sumac
(blooms August)

Smooth Sumac
(blooms July — August)

male (staminate) flowers

Staghorn Sumac
(blooms June — July)

female (pistillate) flowers

Sumac fruit all red
except Poison Sumac
which is whitish

SCALE
ACTUAL SIZE

Staghorn Sumac

Smooth Sumac

Dwarf Sumac

Poison Sumac
fruit develops below the leaves
(fruit of other Sumacs develops
at ends of branches and above
the leaves)

MP
92

Distinguishing Characteristics of Sumac

Poison Sumac

Beware of this one. It's worse than Poison Ivy.

Flower, fruit and leaf unlike other Sumacs.
Twigs are unlike all but Dwarf Sumac, but leaf scars are larger than Dwarf Sumac's.
Leaf has smooth margins — straight stem with no wings along it.

Dwarf Sumac

Flower and fruit unlike Poison Sumac.
Leaf has smooth margins but there are usually wings along the stem.

Smooth Sumac

Fruit similar to Dwarf Sumac but lasts longer.
Twigs smooth, larger than those of Dwarf or Poison Sumac.
Leaflets have teeth along the margins, very similar to Staghorn Sumac leaves.

Staghorn Sumac

Fruit fuzzy, berries indistinct.
Twigs very fuzzy at all seasons.
Leaflets have teeth along margins.

Staghorn Sumac
(smooth Sumac similar)

SCALE
½ ACTUAL SIZE

Dwarf Sumac
(notice wings along leaf stem)

Poison Sumac

MP 93

AMERICAN HOLLY – *Ilex opaca*

Flowers Dioecious—June
Height medium
Southern part of range, hardy north to southern New England.

Leaves, flowers, berries
SCALE
ACTUAL SIZE

The only broad-leaved tree evergreen in northern part of range

inconspicuous flowers

conspicuous red berries

MP 94

AMERICAN LINDEN – *Tilia americana*
(Basswood or Lime)

This column
SCALE
ACTUAL SIZE

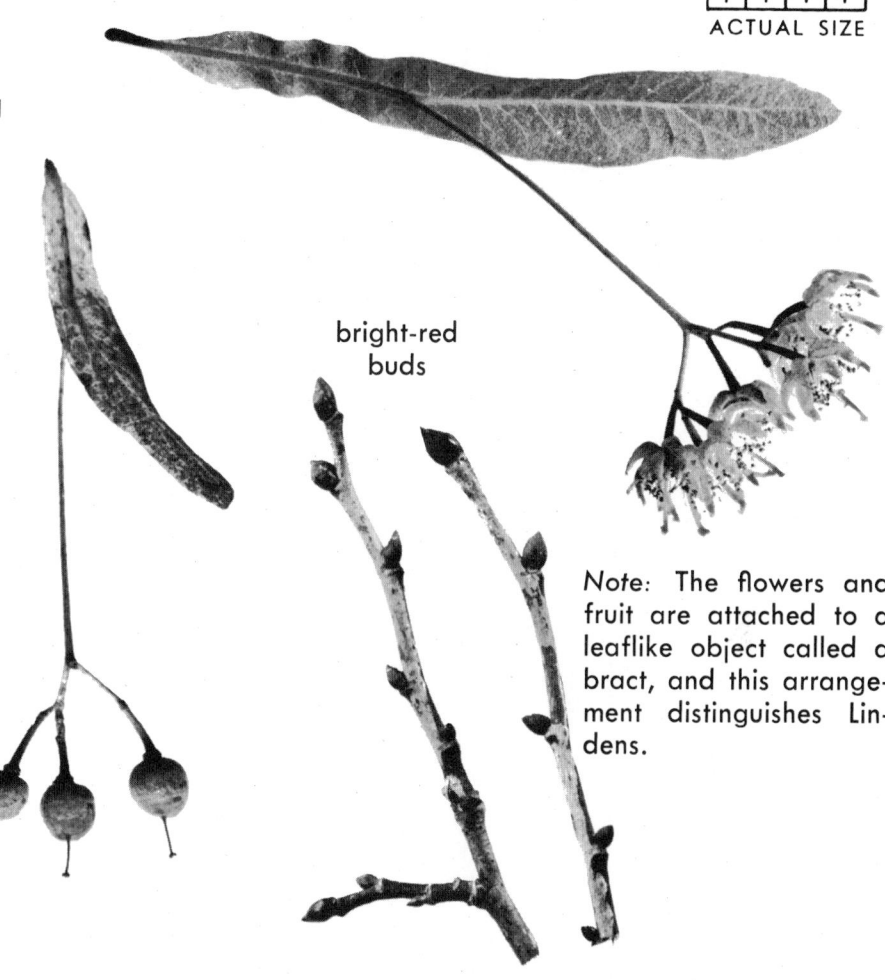

Flowers Perfect—July
Large
Northern half of range extending south in Appalachians.

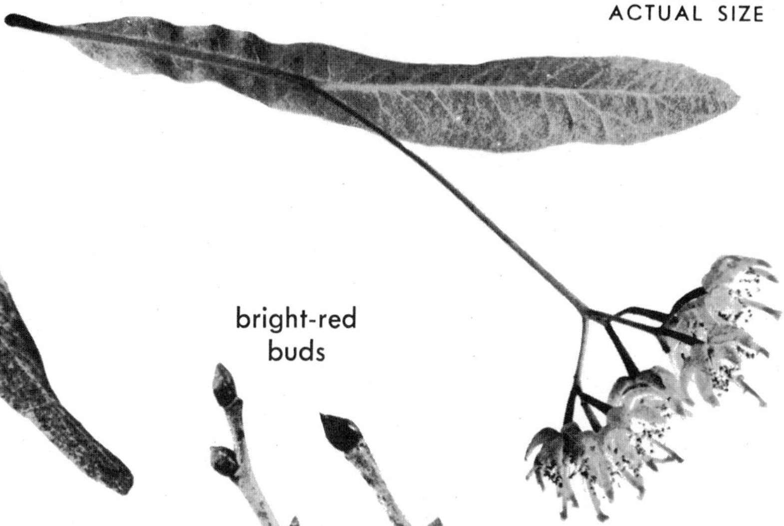

Note: The flowers and fruit are attached to a leaflike object called a bract, and this arrangement distinguishes Lindens.

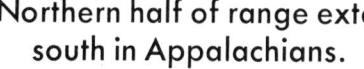

MAPLE – *Acer* Flowers Monoecious or Dioecious—March-May (see individual flowers)

Ash-leaved Maple (Box elder)	*Acer negundo*	medium	Most of range
Mountain Maple	*A. spicatum*	small	Northern states and Appalachians
Norway Maple	*A. platanoides*	large	From Europe, now in most of range

Norway Maple

Silver Maple

Sycamore Maple

Ash-leaved Maple

Red Maple (Swamp Maple)	*A. rubram*	large	Most of range
Silver Maple	*A. saccharinum*	large	Most of range
Striped Maple (Moosewood)	*A. pennsylvanicum*	small	Northern states and Appalachians
Sugar Maple (Rock Maple)	*A. saccharum*	large	Most of range except extreme South
Sycamore Maple	*A. pseudoplatanus*	large	From Europe. Hardy except extreme North.

(old) Sugar Maple (young)

(old) Red Maple (young)

SCALE
½ ACTUAL SIZE

Maple leaves

Note: Norway Maple leaves are broader in relation to their length than those of Sugar Maple.

Sugar Maple

Norway Maple

Striped Maple
(usually very large)

Sycamore Maple

Maple leaves

SCALE
½ ACTUAL SIZE

Silver Maple

Red Maple

Mountain Maple

usually three leaflets but
sometimes five, as shown

Ash-leaved Maple
(Box Elder)

SCALE
ACTUAL SIZE

Silver Maple
(blooms March)

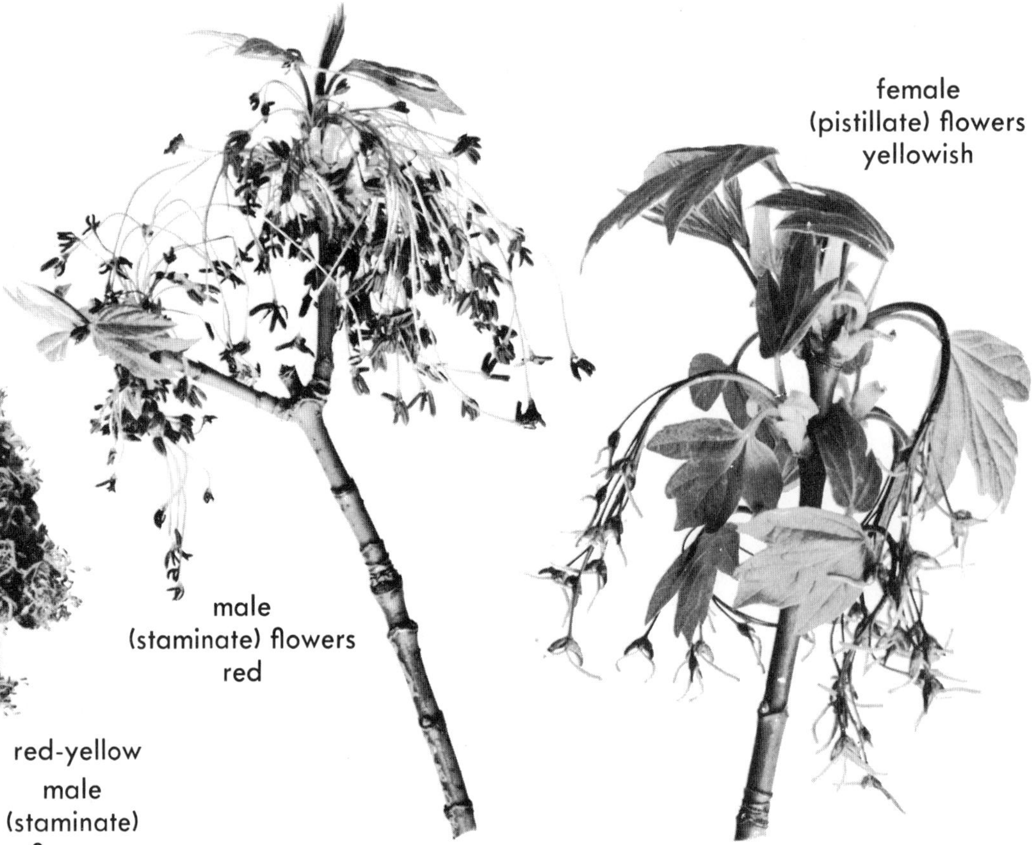

Ash-leaved Maple
(blooms April — May)

Red Maple
(blooms March — April)

Striped Maple
(yellow — blooms May)

SCALE
ACTUAL SIZE

Norway Maple
(yellow — blooms May)

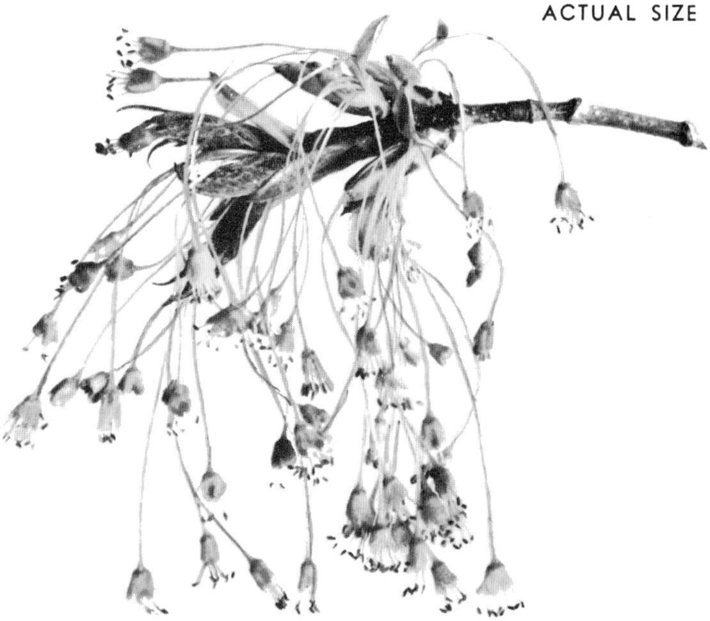

Sugar Maple
(yellow — blooms May)

Sycamore Maple
(yellowish — blooms May-June)

Mountain Maple
(yellowish — blooms May-June)

SCALE
ACTUAL SIZE

Maple fruit

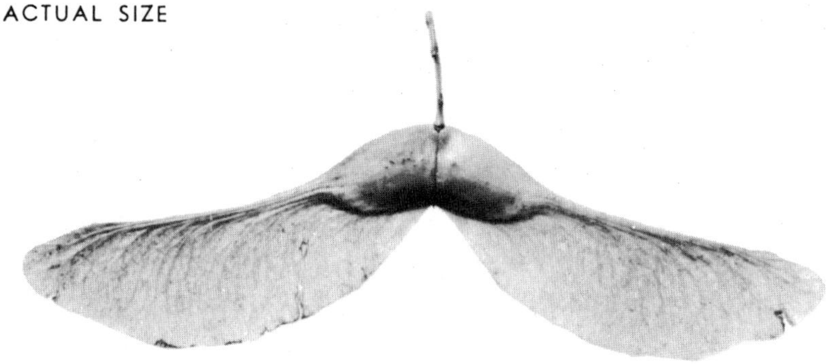
Norway Maple

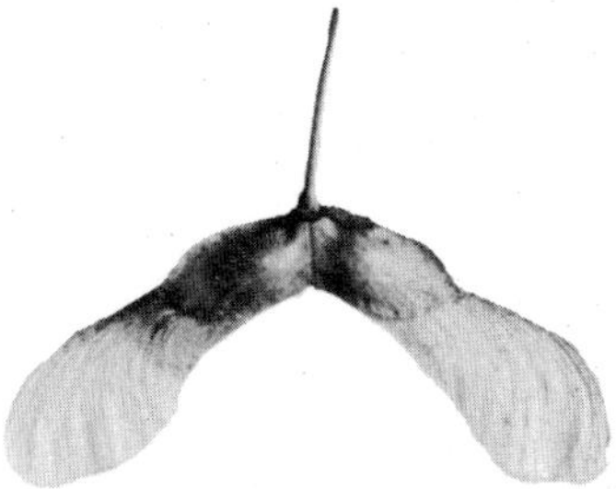
Striped Maple

Sycamore Maple

Sugar Maple

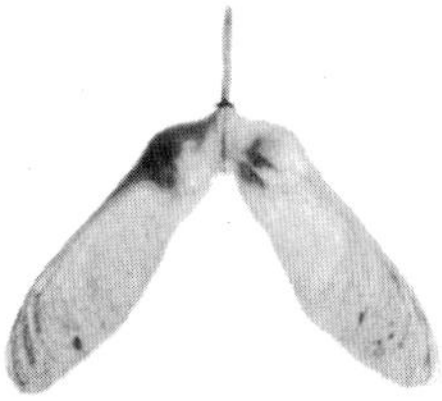
Mountain Maple

Ash-leaved Maple

Silver Maple

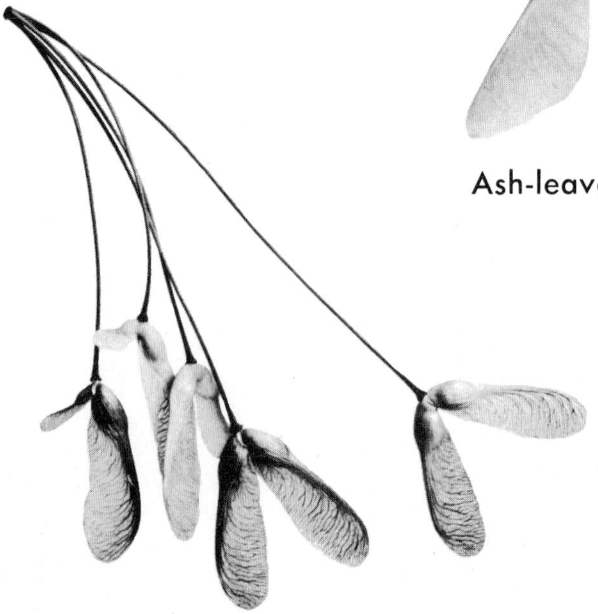
Red Maple

Maple twigs—note opposite buds

SCALE
ACTUAL SIZE

red buds and twigs | brown buds and twigs | red buds brown twigs | brown buds and twigs | often two buds at end of twig greenish buds brown twigs

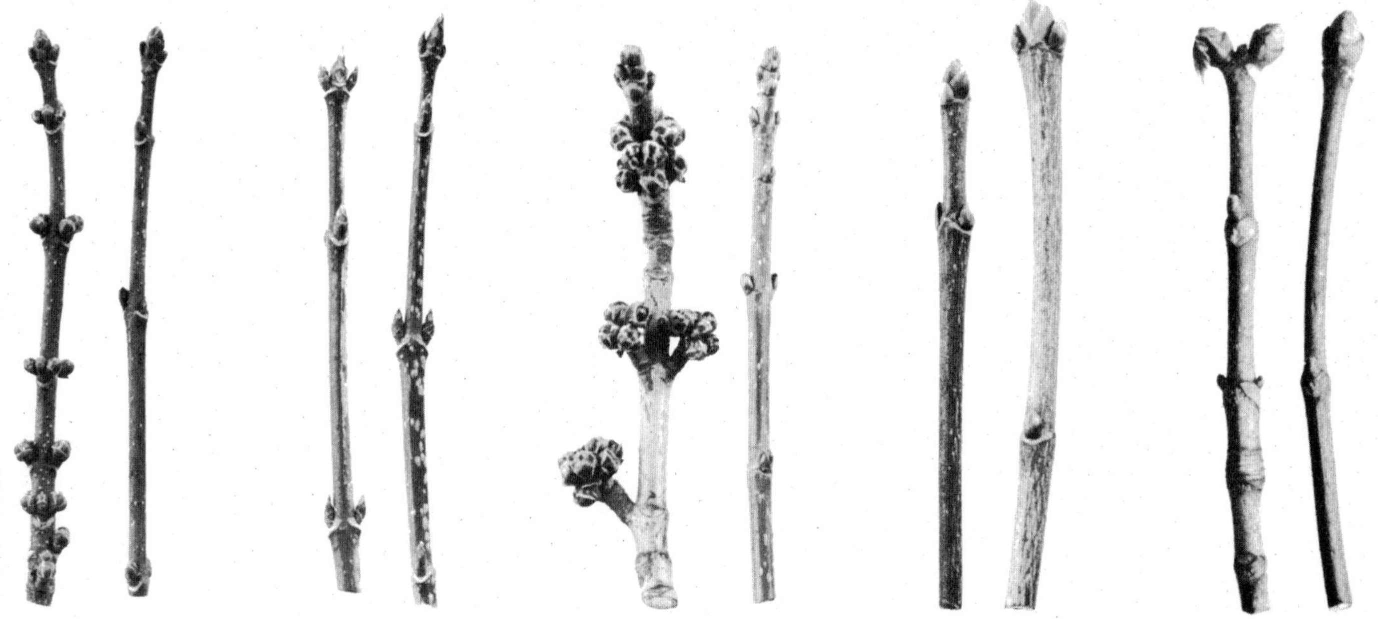

Red Maple | Sugar Maple | Silver Maple | Norway Maple | Sycamore Maple

olive-green twigs, light-colored buds

red twigs and buds, twig sometimes is yellow or green underneath

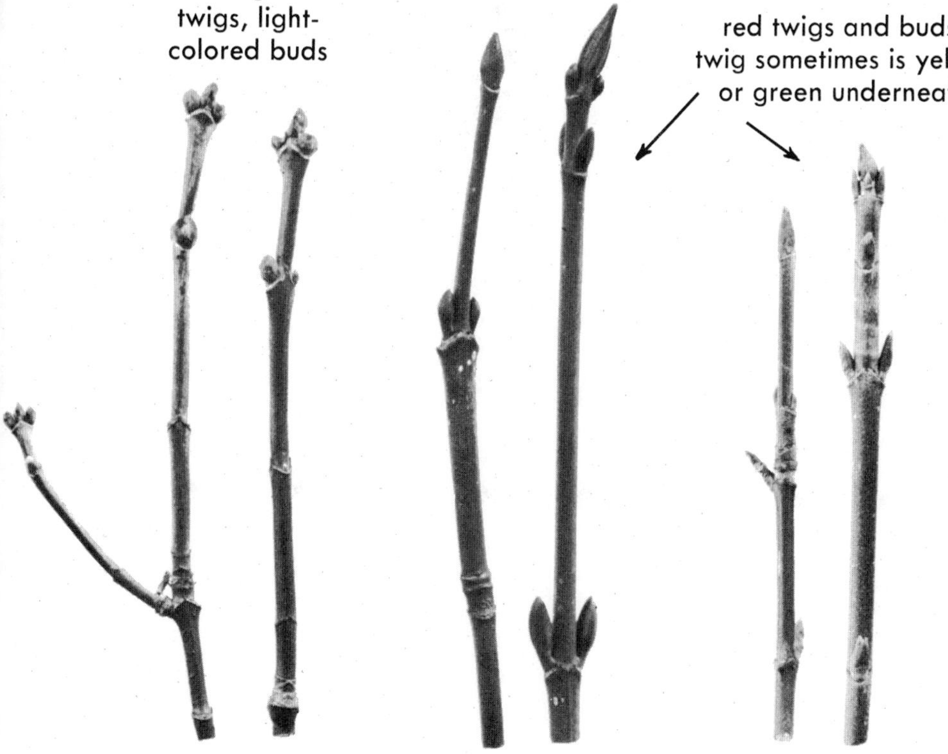

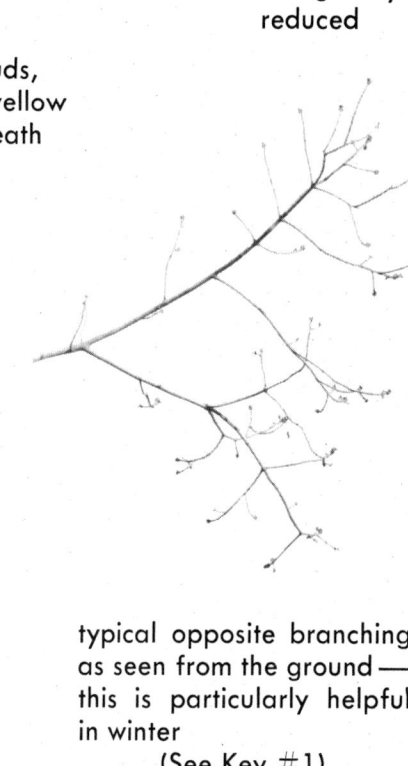

Ash-leaved Maple | Striped Maple | Mountain Maple

branch greatly reduced

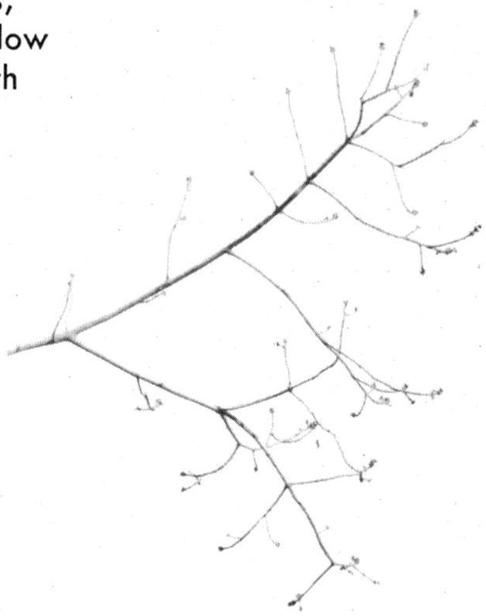

typical opposite branching as seen from the ground — this is particularly helpful in winter

(See Key #1)

Red Maple

Sugar Maple

Red Maple bark usually does not change much after it reaches this size, but the scales of Sugar Maple bark tend to enlarge with age.

Red Maple

Sugar Maple

Red Maple—(young bark)

Although larger than tree above, this is younger.

Sugar Maple (young bark)

Norway Maple

Ash-leaved Maple

Silver Maple

Sycamore Maple

SCALE
ACTUAL SIZE

Striped Maple is almost always a small tree, and this type of bark, green with white stripes, is very characteristic.

Striped Maple

BUCKEYE HORSECHESTNUT — *Aesculus*

Flowers Perfect or Monoecious—May

Horsechestnut	*Aesculus hippocastanum*	medium to large	From Asia via Europe, now in most of range
Ohio Buckeye	*A. glabra*	medium	Found mostly central part of range
Sweet Buckeye	*A. octandra*	medium to large	Usually more easterly but overlapping Ohio Buckeye's range

Ohio Buckeye

Sweet Buckeye

Ohio Buckeye

Sweet Buckeye

This column
SCALE
ACTUAL SIZE

notice descending branches turning up at the ends, typical of Horsechestnut

Horsechestnut

buds very sticky and shiny

Horsechestnut

buds *not* sticky, smooth, points of scales close to bud (appressed)

Sweet Buckeye

buds *not* sticky, points of bud scales tend to stick out (divergent)

Ohio Buckeye

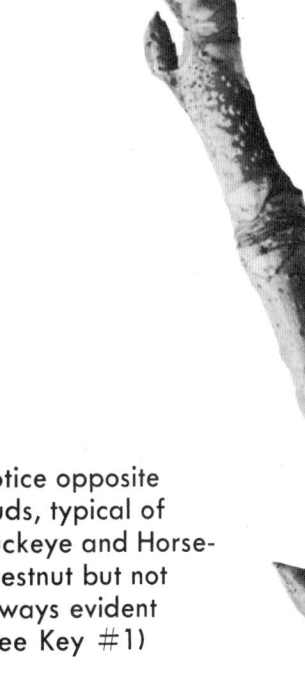

opposite bud missing here which will cause alternate effect when new twig grows from single bud

notice opposite buds, typical of Buckeye and Horsechestnut but not always evident (See Key #1)

Sweet Buckeye

Horsechestnut

SCALE

½ ACTUAL SIZE

Note: The stamens of the Ohio Buckeye and Horsechestnut flowers are clearly evident, extending well beyond the petals, while the stamens of the Sweet Buckeye flowers are not apparent. This is an important feature to notice.

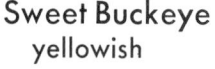

Sweet Buckeye
yellowish

Ohio Buckeye
yellowish

Horsechestnut
white flecked with red

MP 108

Fruit
SCALE
ACTUAL SIZE
Sweet Buckeye
thick, smooth husk
Ohio Buckeye
thin, prickly husk
Horsechestnut

Leaves
SCALE
1/3 ACTUAL SIZE
Buckeyes
usually
five leaflets
Horsechestnut
usually
seven leaflets

MP 109

TUPELO – *Nyssa sylvatica*

(Black Tupelo, Black Gum, Sour Gum, Pepperidge, Beetle Bung Tree)

Flowers Dioecious (sometimes Perfect)—May-June
Medium to large
Most of range except northwest section

Tupelo leaves often grow in clusters and vary from short and wide to long, narrow and tapering. They are usually dark, shiny green, but are about the first to turn color in fall and are a brilliant red, while most trees are still green.

The Tupelo has a very twiggy look, an aid to recognition in winter.

male (staminate) flowers

MP 110

The bark varies from knobbly and deeply furrowed through an intermediate type

female
(pistillate)
flowers —
very often in
threes, resulting in
clusters of three
berries

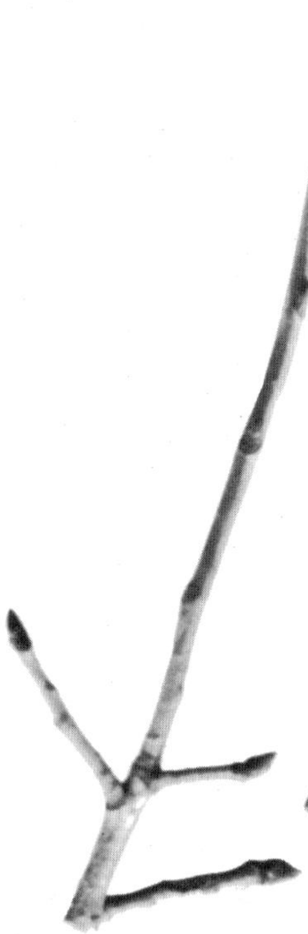

Leaves, flowers, fruit, twigs

SCALE

ACTUAL SIZE

The small branches have smooth light-gray bark, sometimes tinged with pink. The twigs are reddish, sometimes light gray like the branches.

blue-black
berries

to relatively smooth bark. Sometimes more than one type on the same tree.

DOGWOOD – *Cornus*

Flowers Perfect—May-June

Alternate-leaved Dogwood	*Cornus alternifolia*	small	Northern part of range, extending down Appalachians
Flowering Dogwood	*C. florida*	small	Most of range except extreme North

(Pink Dogwood, *C. f. rubra,* is a variety of the White Flowering Dogwood commonly sold in nurseries.)

Note: The Japanese Dogwood, *C. kousa,* is a small tree sold by nurseries. It somewhat resembles our native Flowering Dogwood, but has smoother bark and blooms about a month later. (See Flower 22.)

There are many Dogwoods, ranging from tiny plants only a few inches high to small trees like the two shown here. They all have opposite leaves and branches except the Alternate-leaved Dogwood, a variation so unusual that both the popular and botanical names are based on it. The opposite growth characteristics of the others are always pronounced and very helpful for purposes of identification.

Flowering Dogwood

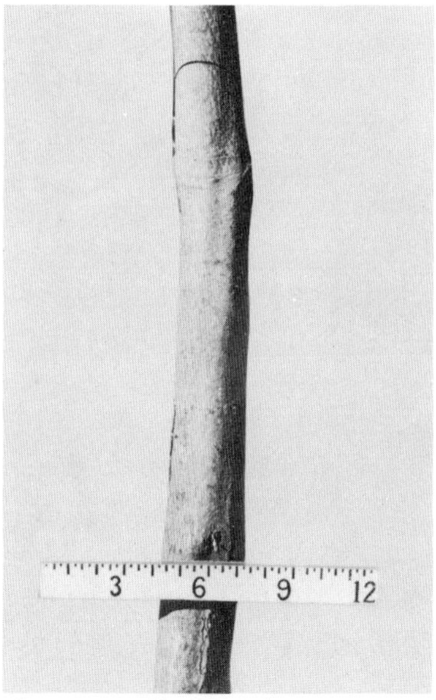

Alternate-leaved Dogwood

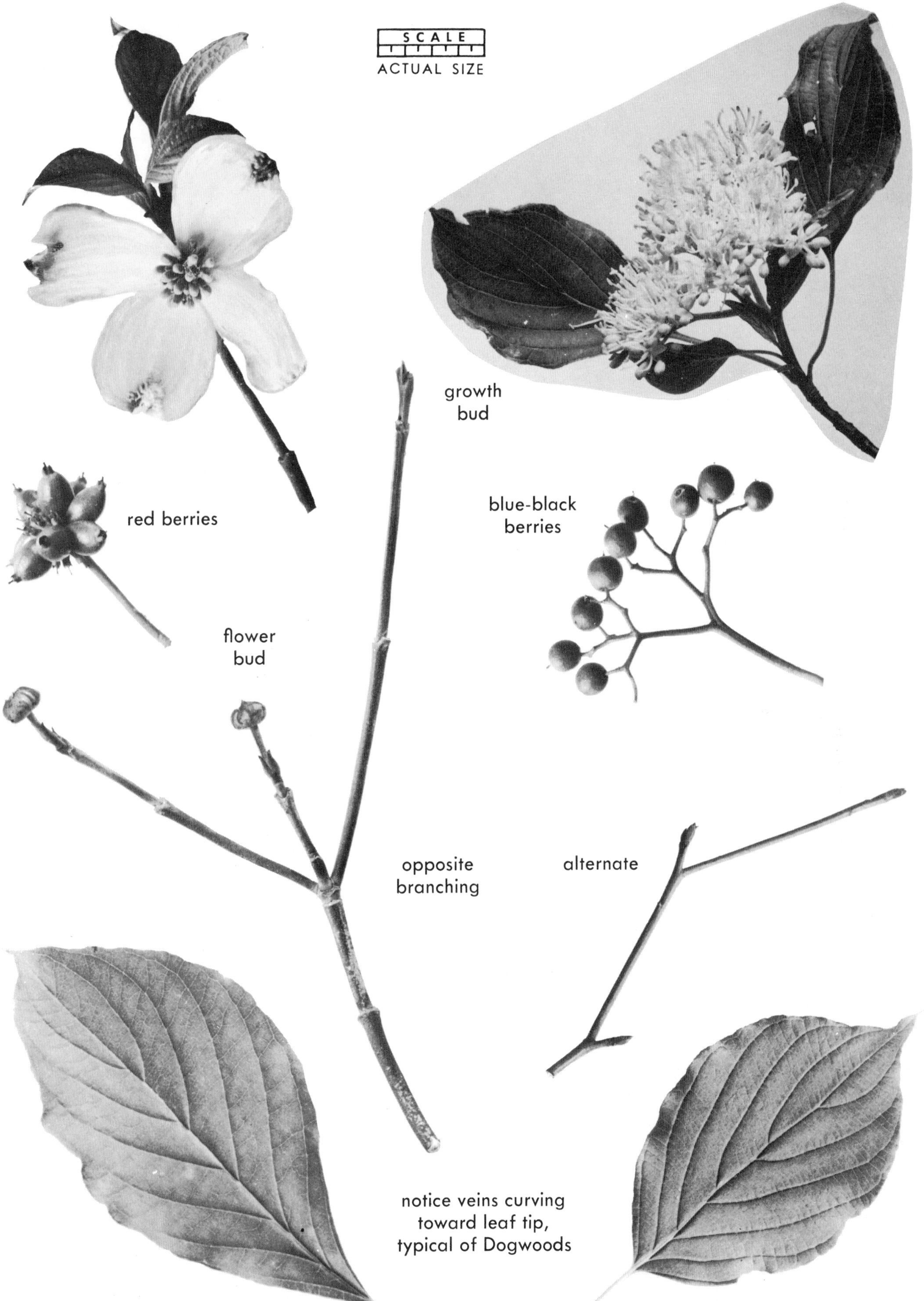

Flowering Dogwood

Alternate-leaved Dogwood

SOURWOOD – *Oxydendrum arboreum*
(Sorrel Tree)

This column
SCALE
ACTUAL SIZE

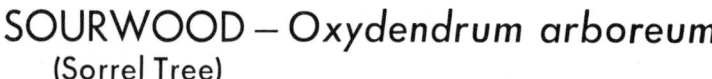

Flowers Perfect—July-August
Medium
Mostly southeastern states, but hardy to southern New England

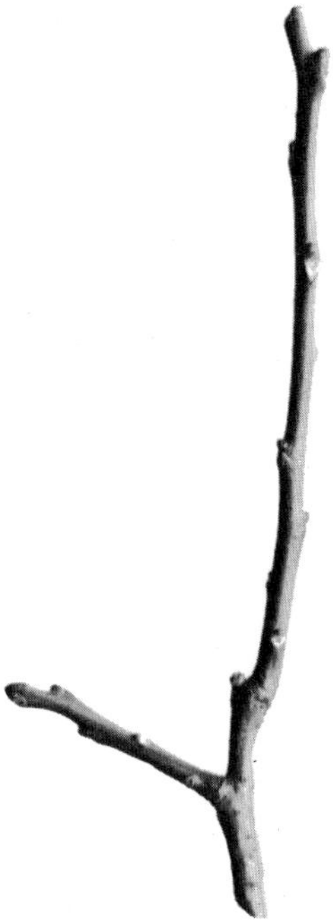

flowers

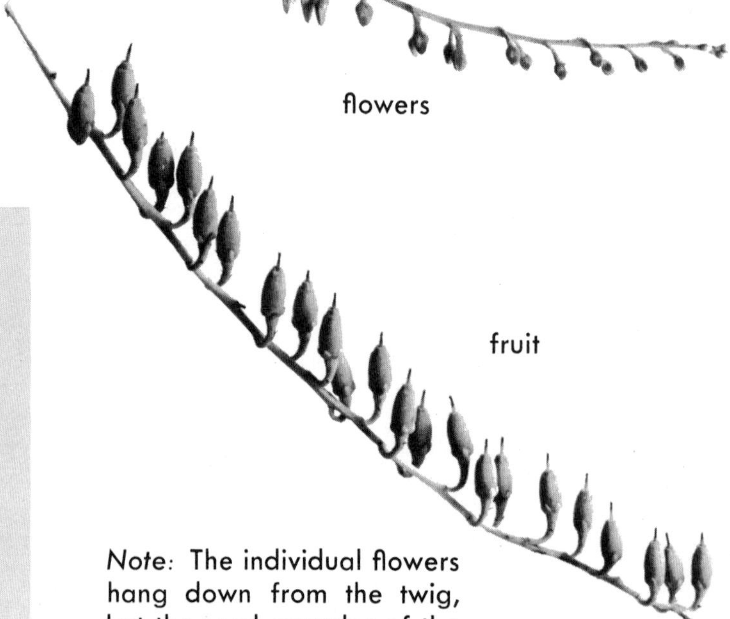

fruit

Note: The individual flowers hang down from the twig, but the seed capsules of the fruit turn up.

MP
114

PERSIMMON – *Diospyros virginiana*

Flowers Dioecious—June
Medium to large
Southern and north to southern
New England and Nebraska

male (staminate)
flowers
whitish

female (pistillate)
flowers
light yellow

This column
SCALE
ACTUAL SIZE

MP
115

ASH – *Fraxinus* Flowers usually Dioecious except Blue Ash which has Perfect flowers—May

Black Ash	*Fraxinus nigra*	medium to large	Found mostly in swamps—northern
Blue Ash	*F. quadrangulata*	medium to large	Mostly central part of range
Red Ash	*F. pennsylvanica*	medium to large	Most of range
White Ash	*F. americana*	large	Most of range

The Green Ash, *F. p. lanceolata* is considered a variety of the Red Ash, the main difference being hairy twigs and leaves on the Red Ash and a lack of these hairs on the Green Ash. There are intermediate forms, however, which makes positive identification difficult.

Black Ash

Blue Ash

Black Ash

Blue Ash

MP
116

Ashes tend to cross (or hybridize) among themselves which makes species identification confusing at times. However, when the details remain true, identification of these four Ashes is not difficult.

Red Ash

White Ash

Red Ash

White Ash

MP 117

Typical opposite branching

This is a particularly helpful feature in winter when looking up into a tree. Compare the few large twigs of the Ash with the small, numerous and also opposite twigs of the Maple (see M.P. 103.)

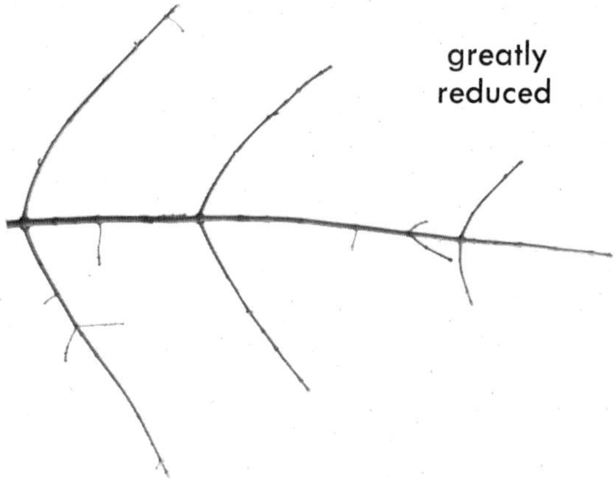

greatly reduced

typical opposite buds

SCALE

ACTUAL SIZE

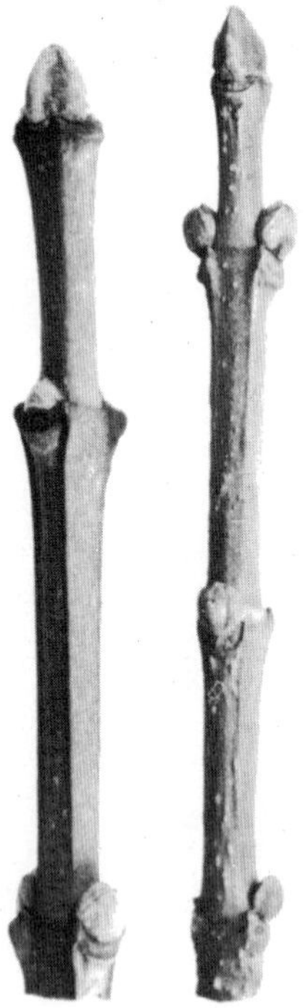

Blue Ash

Notice four-sided twigs — the Blue Ash derives the botanical name "quadrangulata" from these twigs, but they are not always obviously so. The inner bark produces a blue dye which accounts for the common name. This can be tested by breaking the twigs in water.

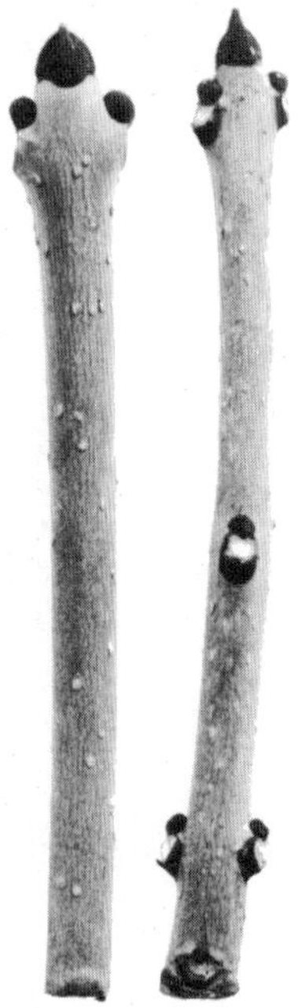

Black Ash

Usually the buds of Black Ash are darker than those of other Ashes.

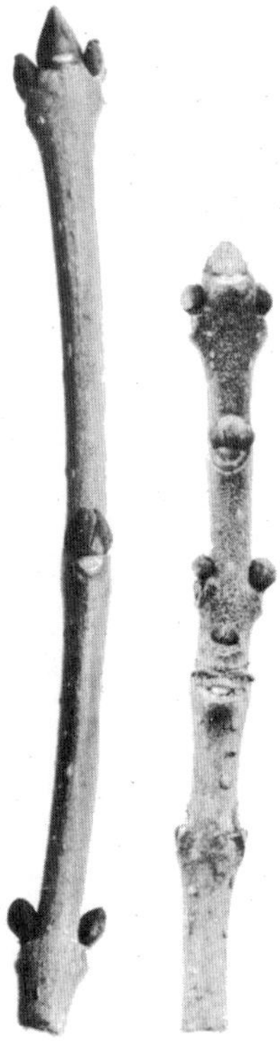

Red Ash

A truly typical Red Ash twig is very fuzzy, but there are intermediate forms which tend toward the hairless twigs of the Green Ash, in which case hybridization would be suspected.

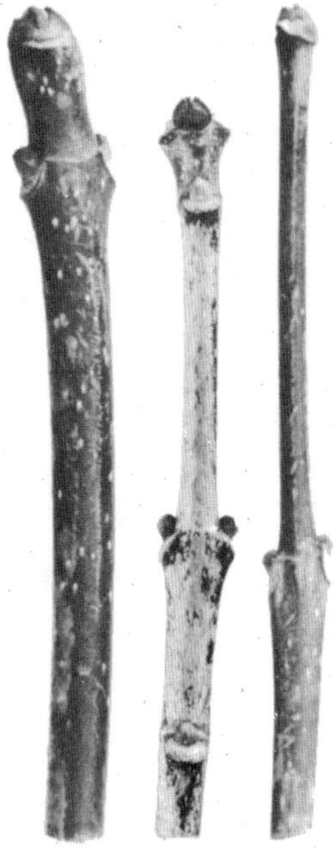

White Ash

Smooth, hairless twigs; usually stubby buds

SCALE
ACTUAL SIZE

typical Ash flowers

Male Ash flowers malformed by disease or insects usually hang on most of winter and are, therefore, very conspicuous.

Blue Ash
perfect
flowers

Blue Ash | Black Ash | Red (& Green) Ash | White Ash

broad — narrow

Other Ashes occasionally are monoecious, or have perfect flowers, but usually are dioecious.
Blue Ash usually blooms before other Ashes.

Note: The seeds of the Red (or Green) Ash are nearly half the total length, while those of the White Ash are much shorter, about one-third.

SCALE

½ ACTUAL SIZE

White Ash

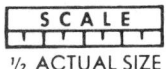

The White Ash leaflet usually is attached to the main leaf stem by a fairly long stem of its own.

Red Ash

The leaves of the Red Ash are hairy underneath. The twigs also are fuzzy. This feature distinguishes the Red Ash from its variety, the Green Ash. This is more easily seen with a magnifying glass, but a good example can be seen by the eye or felt by the hand.

SCALE
½ ACTUAL SIZE

Black Ash

The Black Ash leaflet has practically no stem of its own and is attached close to the main stem of the leaf. The leaflets of other Ashes almost always have short stems.

Blue Ash

The Blue Ash almost always has distinct teeth along the margins of the leaflets. The others sometimes do but often do not. The leaflets of Blue Ash generally are narrower than those of other Ashes.

PAULOWNIA – *Paulownia tomentosa*
(Royal Paulownia, Empress or Princess Tree)

This column
SCALE
ACTUAL SIZE

Flowers Perfect—May
Medium to large
From China and Japan—extensively planted in South, but found as far north as southern New England

notice opposite leaf scars

these are flower buds

lavender flowers

Fruit
SCALE
ACTUAL SIZE

mature fruit pods open to disperse tiny winged seeds

immature fruit

Leaves, flowers,
SCALE
½ ACTUAL SIZE

Very large leaves grow opposite each other along the twigs. Two typical forms shown here.

MP 123

CATALPA – *Catalpa* Flowers Perfect—June-July

Common (or Eastern) Catalpa	*Catalpa bignonioides*	medium	Southern tree naturalized in much of North
Western (or Northern) Catalpa	C. *speciosa*	large	Originaly middle West, now generally planted in north-eastern states

Common Catalpa
(broad, spreading growth)

Western Catalpa
(tall, narrow growth)

Common Catalpa
(thin, loose scales)

Western Catalpa
(coarse, thick scales)

MP
124

The Common Catalpa is best distinguished from the closely related Western Catalpa by its manner of growth and its bark (see opposite page).

The Common Catalpa blooms two weeks later than the Western one, when growing in the same area.

The pod of the Common Catalpa is generally narrower and the shell thinner than that of the Western Catalpa. Individual pods of both species vary greatly, so that species identification should not be based on pods alone.

SCALE
ACTUAL SIZE

The pods crack when mature to disperse winged seeds with fringes of hairs at each end.

MP
125

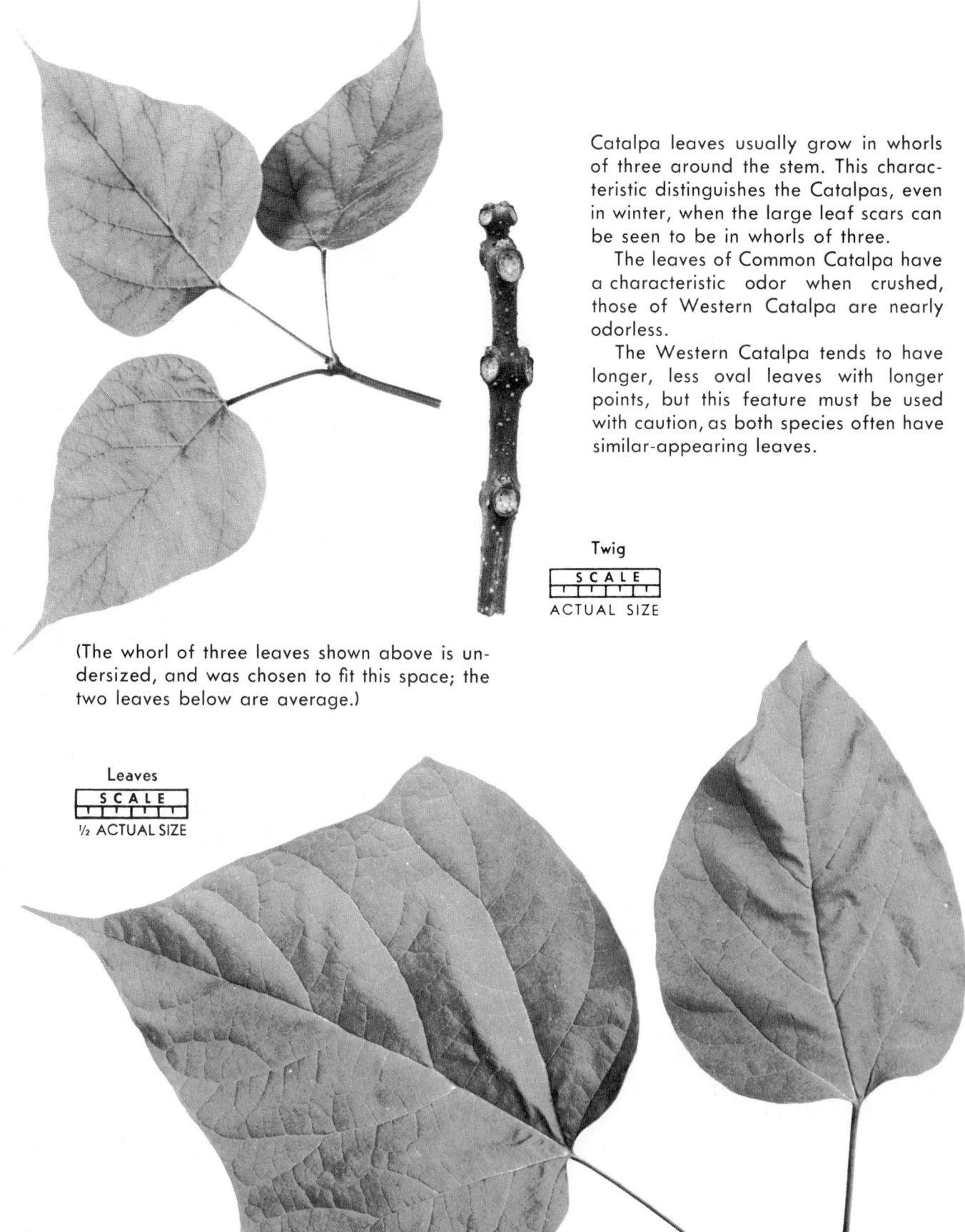

Catalpa leaves usually grow in whorls of three around the stem. This characteristic distinguishes the Catalpas, even in winter, when the large leaf scars can be seen to be in whorls of three.

The leaves of Common Catalpa have a characteristic odor when crushed, those of Western Catalpa are nearly odorless.

The Western Catalpa tends to have longer, less oval leaves with longer points, but this feature must be used with caution, as both species often have similar-appearing leaves.

Twig

SCALE

ACTUAL SIZE

(The whorl of three leaves shown above is undersized, and was chosen to fit this space; the two leaves below are average.)

Leaves

SCALE

½ ACTUAL SIZE

MP 126

NANNYBERRY—*Viburnum lentago*
(Sheepberry or Sweet Viburnum)

This column
SCALE
ACTUAL SIZE

Flowers Perfect—June
Small, often shrublike
Most of range

There are a great many Viburnums, mostly shrubs, but a few occasionally take tree form.

berries turn
from white
to blue-black

distinctive
end bud

Viburnums have definite opposite growth characteristics.
Nannyberry is sometimes mistaken for Dogwood, but the combination of details is unlike any Dogwood.

MP
127

APPENDIX

Introduction to Botanical Manuals and Methods

There are a number of good botanical manuals on the market; however, Alfred Rehder's *Manual of Cultivated Trees and Shrubs,* second edition, published by Macmillan, has been selected to illustrate what follows. If this book is not available at your local library, try to obtain a substitute recommended by the librarian. In any event, it will be necessary to have a botanical manual of some sort. Some knowledge of botanical terms is essential. The glossary included in most manuals will be helpful. A magnifying glass, preferably 12-15 power will also be needed.

Before undertaking to pursue the subject further, it should be understood that words do not always express an exact meaning. For example, "Pubescent" is described as, "Covered with hairs, particularly if short and soft." "Tomentose" means, "With dense woolly pubescence." Now, this is all very well, but the exact difference is not clear, and what one man may believe is pubescent, another may call tomentose. And here's the rub: a key based on the assumption that words alone can express an exact meaning, and therefore can be interpreted precisely as intended, may be misleading, unless the reader knows enough of the author's work to be able to decide how a particular word should be interpreted. This may seem farfetched, but is probably the main stumbling block in following a botanical key. Therefore, take heed of this, and much disappointment will be avoided. Don't be afraid to backtrack in a given key if it seems to be leading you in a wrong direction.

In addition to the divisions in the plant kingdom used in this book, namely genera and species, there are others, both larger and smaller. The group larger than genus is called a family. Families are grouped together into larger divisions known as orders, classes and phyla. The phylum is the largest group of related plants. It will be noted that botanical manuals are arranged according to these divisions, and within the larger divisions the families are the leading headings. Then in the different families will be found the various genera, and then, of course, our old friend the species. Now, this may seem confusing, but don't let complicated-sounding words deter you. It all makes very good sense, and instead of trying to learn a lot of new terms at once, just sample them and then go right to the back of the book. There in the index will be found things that you know. For example, if by using our Pictorial Keys you decide that a tree is some obscure member of the genus Oak, turn to the index in the manual and look for *Oak*. This will refer you to page 153 (in Rehder), and on that page will be found "Quercus L. Oak." The L. stands for Linnaeus, short for Carl von Linné, the Swede who started the whole business of scientific nomenclature and proper classification. Under the Oak section will be found a key, and you will see that the listing, though long, is not hopeless, and that it is not too difficult to separate the true from the false. Immediately following the key in numerical order, the numbers corresponding to those indicated in the key, is a short description of the various Oaks, this time giving the English names as well as the botanical ones.

If you can't tell the genus of an unknown tree, study a little about families, the next bigger group. If you can determine the family, you will have reduced the field considerably. From this point, by using the family key in a manual, you will undoubtedly reduce it to the genus, and then you will find the going good again.

Do not expect 100 per cent success, and in some cases you would do well to consult a local nurseryman. If he can identify an unknown tree for you, study it and compare it to the description in the manual. Another good procedure is to study trees in a nursery, as they will have a number of foreign trees or those from another region, which, although not covered in this book, will be found in Rehder.

BIBLIOGRAPHY

BAILEY, LIBERTY H. *The Standard Cyclopedia of Horticulture.* 3 vols. New York: Macmillan Co., 1935.

COLLINGWOOD, G. H. and BRUSH, W. D. *Knowing Your Trees.* Washington: The American Forestry Assoc., 1955.

CURTIS, C. C. and BAUSOR, S. C. *The Complete Guide to North American Trees.* New York: The New Home Library, 1943.

EMERSON, ARTHUR I. and WEED, CLARENCE M. *Our Trees: How to Know Them.* Philadelphia: J. B. Lippincott Co., 1937.

FERNALD, M. L. *Gray's Manual of Botany.* 8th ed. New York: American Book Co., 1950.

GLEASON, HENRY A. *The New Britton and Brown Illustrated Flora.* 3 vols. New York: New York Botanical Garden, 1952.

GRAVES, ARTHUR HARMOUNT. *Illustrated Guide to Trees and Shrubs.* Revised ed. New York: Harper & Brothers, 1956.

KEELER, HARRIET L. *Our Native Trees.* New York: Charles Scribner's Sons, 1900.

KIERAN, JOHN. *Introduction to Trees.* Garden City: Hanover House, 1954.

REHDER, ALFRED. *Manual of Cultivated Trees and Shrubs Hardy in North America.* 2nd ed. New York: Macmillan Co., 1943.

SARGENT, CHARLES S. *Manual of the Trees of North America.* Boston: Houghton Mifflin Co., 1933.

———. *The Silva of North America.* 14 vols. Boston: Houghton Mifflin Co., 1891-1902.

Catalogue of Hardy Trees and Shrubs. New York: New York Botanical Garden, 1943.

GLOSSARY

Appressed—close to (not sticking out); opposite of divergent.

Bract—a modified leaf, usually at the base of flowers or leaves, sometimes seen protruding beyond the scales of cones.

Compound—having two or more similar parts; a leaf made up of several leaflets on the same stem is a compound leaf.

Deciduous—not permanent, falling off, as with leaves in autumn.

Dioecious—male flower on one plant, female on another.

Divergent—sticking out; opposite of appressed.

Genus—main name of a plant (Oak, Maple, etc.).

Hybrid—a cross between closely related plants.

Lateral—situated or coming from the sides; lateral buds are those along a twig, as distinguished from end or terminal buds.

Leaflets—the single blades of a compound leaf.

Leaf scar—the scar left on a twig, marking the place the leaf was attached before falling.

Lobe—the part of a leaf sticking out when the margins are not uniform. This is the opposite of the sinus, which is the cut-out part of an uneven leaf.

Monoecious—male and female flowers distinct from one another, but on the same plant.

Node—the point on the twig which bears a leaf or leaves.

Palmate—spreading from one point; a palmately compound leaf is one where all the leaflets branch from one point.

Perfect—both male and female parts in the same flower.

Pinnate—having parts arranged along two sides; a pinnately compound leaf has leaflets along both sides of the leaf stem.

Pistil—seed-bearing part of a flower—female part.

Pistillate flower—female flower.

Pith—inner part of twig which is not woody but pulpy.

Sessile—without a stem.

Simple—applied to a leaf with only one blade.

Sinus—the cut-out part of a leaf with uneven margins; the opposite of lobe.

Species—first name of a plant, as *Red* Maple, *Black* Oak.

Stamen—the pollen-bearing (male) part of the flower.

Staminate flower—male flower.

Stipules—leaflike objects at the base of leaf stems, not actually part of the real leaf.

Stoma (plural stomata)—a hole or pore on the outside (epidermis) of a leaf.

Whorl—a circle of three or more similar parts around a central point, as three or more leaves growing around a twig at one spot or node.

INDEX